The Manufacturing Guides

Prototyping and Low-volume Production

Rob Thompson

The Manufacturing Guides

Prototyping and Low-volume Production

Thames & Hudson

Contents

Page 2, clockwise from top left: Cathryn Shilling kiln forming glass; mold making with CNC machining; flocking a point-of-sale display; and reaction injection molding an automotive interior.

Copyright © 2011 Rob Thompson

First published in 2011 in paperback in the United States of America by Thames & Hudson Inc., 500 Fifth Avenue, New York, New York 10110

thamesandhudsonusa.com

Library of Congress Catalog Card Number 2010932489

ISBN 978-0-500-28918-1

Printed and bound in China by Toppan Leefung Printing Limited

**Part One
Forming Technology**

How to use this book

This guidebook is intended as a source of inspiration in the creative process. Established and new prototyping, batch production and low-volume manufacturing techniques are covered. The case studies demonstrate the scope for creativity and the process illustrations highlight some of the technical considerations for each of the technologies.

How to use the processes sections

Each process is introduced and a brief outline of the key reasons why it may be selected are put forward. The description focuses on the conventional application of each technology. It is up to the designer and artist to challenge these established ways of working.

The book is divided into three parts (colour coded for ease of reference), which are forming technology (blue), joining technology (orange) and finishing technology (yellow).

Technical illustrations show the inner workings of the processes. These principles are fundamental and define the technical constraints of the tools and equipment. Each technique within a process, such as benchworking (page 83) and latheworking (page 86) which are included in lampworking (page 82), are individually explored and technically explained.

Essential Information

VISUAL QUALITY	●●●●●●●
SPEED	●○○○○○○
MOLD AND JIG COST	●●●○○○○
UNIT COST	●●●●●○○
ENVIRONMENT	●●●●○○○

Related processes include:
• Die Sink EDM
• Wire EDM

Alternative and competing processes include:
• CNC Machining
• Laser Cutting
• Water Jet Cutting

Essential Information
A rough guide to five key features of each process to help inform designers and aid decision-making.

How to use the essential information panels

In addition, the opening spread includes a detailed essential information panel. This defines comparable values for the five key features of each process, which are visual quality, speed, mold and jig cost, unit cost and environmental impacts. The scoring system is relative and based on one point being the lowest and seven points the highest. Of course, the type of product, application and context of use will affect these values. They are intended as rough guidance to help inform designers and aid decision-making (see above).

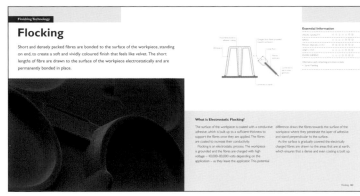

The individual techniques are listed and in many cases explored in more detail in subsequent spreads. Similar processes are often referred to by different names. Where possible, these have been explained and the most fitting process title has been used. For example, 'thermoforming' is used to describe a range of processes used to manufacture high volumes of plastic parts. This group of processes is characterized by the use of air pressure to form sheets of plastic over or into molds. For low volumes, the most common process is vacuum forming (page 22).

How to use case studies

The real-life case studies feature workshops and factories from around the world. They demonstrate the full spectrum of production, from established craft to technical prototyping. Each of these processes is utilized in the development and sometimes full-scale production of well-known products, furniture and artwork.

The processes are covered by a step-by-step description and analysis of the key stages. The principal attributes of each technology are described in detail and some of the additional qualities, such as scale and material scope, are outlined where necessary.

Processes and case studies Each manufacturing technology is described in detail, includes a technical illustration and a case study by a leading manufacturer. This example demonstrates the scope and opportunities of flocking for automotive, interior and furniture applications.

Photographs of geometry, detail, colour and surface finish are used to show the many opportunities that each process has to offer.

Relevant links between the processes, such as forming and finishing operations, are highlighted in the text. It is essential that designers are aware of the wide range of manufacturing opportunities at their disposal. This information provides a well-informed starting point for further focused investigation which is essential for designers to harness the full potential of each manufacturing technology.

Introduction

This book features some of the most inspiring manufacturing and craftsmanship used in the production of products, furniture, prototypes and works of art.

Each technique, ranging from liquid-forming to additive manufacturing, provides designers with a range of opportunities. Projected volumes, geometry, size, performance requirements and aesthetic qualities determine the final selection.

Design for craftsmanship

Prototyping and low-volume production provide a great deal of scope – even for the most humble materials. The choice of material defines the tools, methods and opportunities from the outset. Therefore, knowledge about the material properties and applicable processes are part of a designer's toolbox and are essential for making the right choices.

There are four main categories of forming technology that designers need to be aware of. These include molding and casting; machining and cutting; bending and pressing; and rapid prototyping.

Molding and casting technologies

Liquid-forming techniques can be utilized to shape metals and plastics into complex and intricate parts. They are suitable for making at any size, from tiny scale models (see centrifugal casting, page 128) to bronze sculptures weighing several tonnes (see lost wax casting, page 30).

Plastic is either vacuum cast (page 18) or reaction injection molded (page 14 and see below); used for prototypes, one-offs and low volumes, these manufacturing technologies can replicate almost all the properties of injection molding.

Maverick Television Awards Polyurethane is the most suitable material for vacuum casting and is available in a wide range of densities, colours and hardnesses. It is added to the mold as two separate chemicals, which react to form the plastic part. It can be very soft and flexible or rigid. The Maverick Television Awards were manufactured by CMA Moldform.

Heavy Light Designed by Benjamin Hubert, these hand-cast concrete light shades challenge our perception of industrial materials. They are surprisingly light with a thin wall section. Concrete is available in a range of subdued colours including off-white, beige, blue and black.

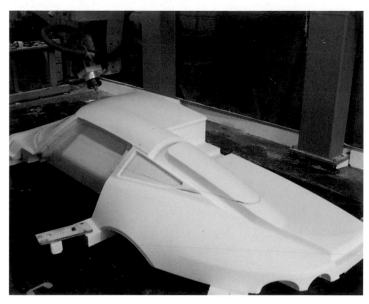

Five-axis CNC machining CNC machines are capable of producing very large parts in a single piece of material such as this full-scale prototype car part.

Many different materials, such as glass and concrete (see above), can be formed in their liquid state using similar techniques. However, the viscosity of these materials will affect the scope of possible geometries and details.

Machining and cutting technologies

Reductive processes include CNC machining (page 42), EDM (page 52), laser cutting (page 112), photochemical machining (page 120), water jet cutting (page 116) and grinding, sanding, polishing (page 170).

Using CNC machining, CAD data can be transferred directly onto the material. The CNC process is carried out on a milling machine, lathe or router, and results in a rapid, precise and high-quality end product. The number of axes that the CNC machine operates on

determines the geometries that can be cut. In other words, a five-axis machine has a wider range of motion than a two-axis machine (see above).

Bending and pressing technologies

This group of processes utilizes the ductility and elasticity of materials. There are many techniques used to bend and press metals (see panel beating, page 38 and press braking, page 124). Composites (see composite laminating, page 94) and wood veneers (see veneer laminating, page 88) are formed at room temperature and cured or kiln fired to finish. Glass, on the other hand, has to be heated to over 600°C (1112°F) for it to become sufficiently soft to take the shape of a mold (see kiln forming glass, page 70).

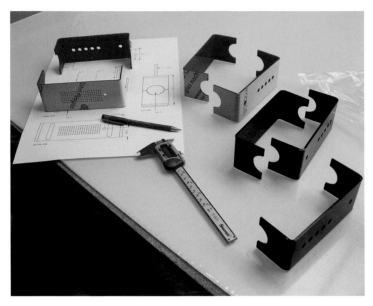

Building a model These parts of a radio prototype were laser cut (page 112) and have been formed by locally heating the plastic and pressing it into a mold. Thermoplastics, such as the poly methyl methacrylate (PMMA) acrylic used here, can be heated and formed relatively easily.

Certain types of plastic can be heated and bent to shape (see above).

Rapid prototyping technologies

These layer-building processes can be used to prototype one-offs or to manufacture low-volume production runs directly from CAD data. There is no tooling involved, which not only helps to reduce the cost, but also has many advantages for the designer. The high tolerances of these processes mean that they are ideal for producing prototypes that are used to test products before committing to full-scale production.

Additional advantages include reducing time to market and lowering product development costs. However, the most desirable quality of this process for designers is not cost savings, but that complex, intricate and previously impossible geometries can be built with these processes to fine tolerances and precise dimensions. There is no tooling and so changes to the design cost nothing to implement. The combination of these qualities provides limitless scope for design exploration and opportunity (see below).

Signature Vase In this case stereolithography (page 110) is being used to direct manufacture people's signatures. Created by Frank Tjepkema for Droog Design in 2003, this idea neatly illustrates the flexibility and speed of this revolutionary technology.

Prototyping for mass production

In the process of design for mass production parts will be prototyped and may be manufactured in low volumes to begin with. Certain manufacturing processes that do not require expensive tooling, such as spray painting (page 180), CNC machining (page 42) and laser cutting (page 112), can be used in a similar way for low to high volumes. It is thus a relatively straightforward transition.

Parts designed for mass production that require very expensive tooling, such as for injection molding and die casting, can be prototyped by reaction injection molding (page 14), vacuum casting (page 18) and rapid prototyping (page 104), for example. These technologies have relatively low tooling costs, but the unit costs are many times higher and the mechanical and visual qualities may be different.

In the absence of mass mechanization, the efficiency and cost effectiveness are less crucial. This provides designers and artists with greater scope to experiment and innovate (see above). Every manufactured item must strike a balance between materials and processes. Designers utilize prototyping and low-volume production to explore the 3D aspects of their ideas. This makes up many stages in the product development process and is fundamental for innovative design.

Studio Glassblowing Each layer of colour in *The Large Stone*, from the Glacier range by London Glassblowing, was created by hand. Studio glass processes are free from the limitations of mass production so there are many ways that glassmakers can manipulate the shape and surface of the glass to create exciting effects.

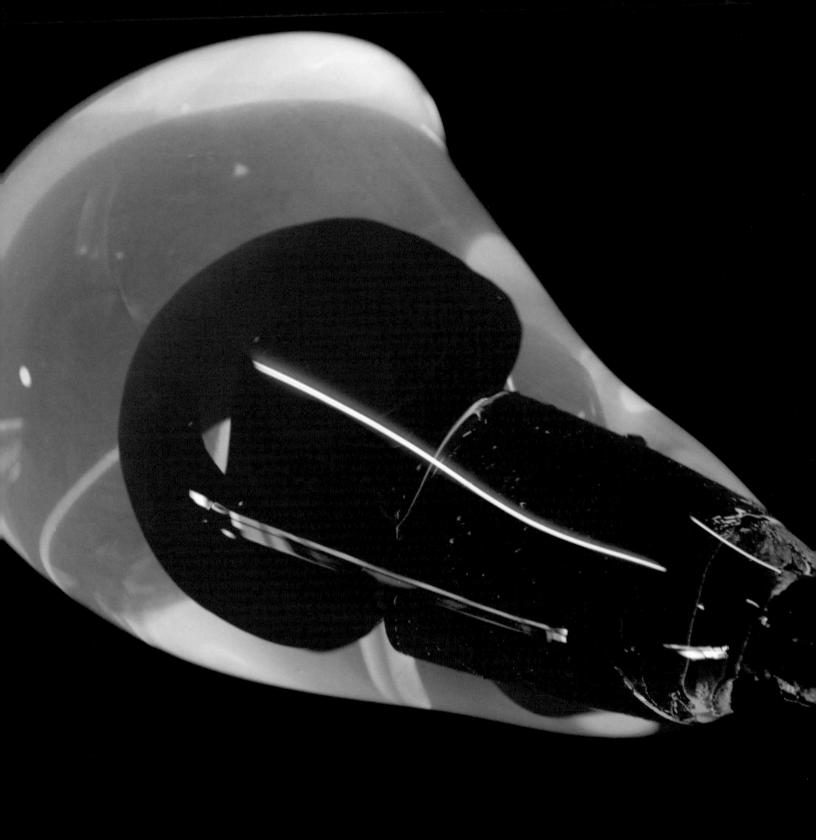

Forming Technology

Reaction Injection Molding

Reaction injection molding is often referred to simply as 'RIM'. The process involves injecting a two-part polyurethane resin (PUR) into a mold at low pressure. Relatively low tooling costs make it ideal for prototyping and low-volume production. It is suitable for making parts up to 2.5 m (8.2 ft) long.

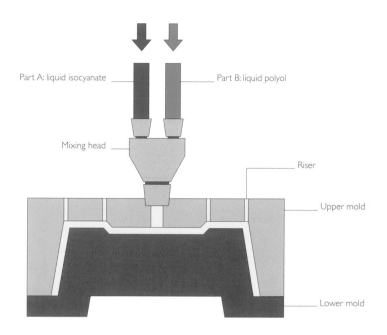

Part A: liquid isocyanate

Part B: liquid polyol

Mixing head

Riser

Upper mold

Lower mold

What is Reaction Injection Molding?

The two ingredients that react to form polyurethane resin (PUR), polyol and isocyanate, are fed into the mixing head where they are combined at high pressure. There are two principal prepolymer systems: toluene diisocyanate based (TDI) and diphenylmethane diisocyanate based (MDI). The predetermined quantities of liquid chemicals are dispensed into the mold at a low pressure (one to two bar). As they are mixed they begin to go through a chemical exothermic reaction. The part is fully cured in about 30 minutes and is demolded.

PUR is a thermosetting resin and the reaction is one way. Therefore, once it has cured, the resin cannot be reused or recycled.

For high strength and more demanding applications the resin can be reinforced with glass fibres, or mica, for example. This is known as reinforced reaction injection molding.

Notes for Designers

QUALITY Even though this is a low-pressure process, the liquid PUR reproduces fine surface textures and details very well. The surface finish is determined by the quality of the mold and spray painting (page 180).

TYPICAL APPLICATIONS Some typical applications include prototypes and low-volume production runs of car bumpers, automotive interior panels, housing for large medical devises and vending machines.

COST AND SPEED Tooling costs are low to moderate. They are considerably less than when tooling for injection molding due to reduced pressure and temperature. Cycle time is quite rapid (15–30 minutes). Labour costs are low to moderate and automated processes reduce labour costs significantly. Prototyping and low-volume production require more labour input.

MATERIALS PUR is the most suitable material because it is available in a range of densities, colours and hardnesses. It can be very soft and flexible or rigid.

ENVIRONMENTAL IMPACTS PUR is a thermosetting material and so cannot be directly recycled. The isocyanates that are off gassed during the reaction are harmful and known to cause asthma. The polyol/MDI system produces fewer isocyanates than the TDI method.

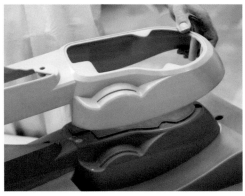

Automotive panel Parts that are finished by spray painting can be masked to highlight design details such as logos and patterns. The level of sheen on the coating is categorized as matt (also known as egg shell), semi-gloss, satin (also known as silk) and gloss. High gloss, intense and colourful finishes are produced by a combination of meticulous surface preparation, basecoat and topcoat.

Complex and intricate details RIM is used to produce parts similar to injection molding. Indeed, it is often used to make prototypes of parts that will be injection molded in the long term because the tooling costs are considerably cheaper, but the unit costs are much higher. Details such as snap fits, threaded inserts and vents can be molded into the part with minimal cost implications.

Prototyping For prototypes and very low volumes RIM molds can be produced by CNC machining blocks of rigid PUR tooling resin. This technique is relatively inexpensive and is a very rapid way of producing a functional prototype suitable for application and testing.

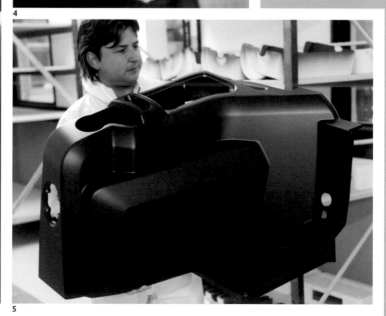

Reaction Injection Molding an Automotive Interior

Featured company Midas Pattern Co. Ltd
www.midas-pattern.co.uk

The cured part is removed from the mold (image **1**). The flash, runners and risers are still intact at this stage. The part is 'fettled' to remove excess material (image **2**).

Once it has been cleaned up the part is carefully abrasively blasted (image **3**) in preparation for spray painting. This ensures that all of the surface residues are removed and it is sufficiently 'keyed' to maximize the strength of the bond between the paint and the PUR. Masking is applied to protect metal inserts (image **4**).

RIM is a low-pressure process and so minor surface defects are inevitable. Spray painting is therefore fundamental for parts that require a high-quality finish. In this case the part is sprayed with a spattered black texture (image **5**), which is ideal for high-wear applications.

Vacuum Casting

Used for prototypes, one-offs and low volumes, vacuum casting can replicate almost all the properties of injection molding. It is primarily used to mold two-part polyurethane (PUR), which is available in a vast range of grades, colours, transparencies and densities.

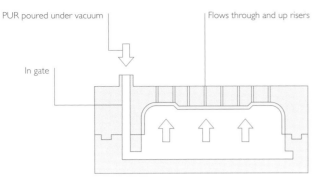

Stage 1: Vacuum casting

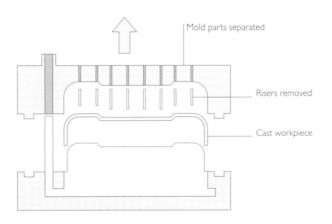

Stage 2: De-molding part

Essential Information

VISUAL QUALITY	●●●●●●○○○
SPEED	●●●○○○○○○
MOLD AND JIG COST	●●●○○○○○○
UNIT COST	●●●●○○○○○
ENVIRONMENT	●●●○○○○○○

Alternative and competing processes include:
- Injection Molding
- Rapid Prototyping
- Reaction Injection Molding

What is Vacuum Casting?

In stage one, the liquid PUR is drawn into the mold which is held under vacuum. This ensures that the material has no porosity, will flow through the mold cavity and will not be restricted by air pressure. The PUR is held high above the mold, so as it is poured into the mold and runs underneath gravity forces it to rise upwards. The runner is designed to supply a curtain of liquid PUR. It flows through the mold and up the risers in part A. There are lots of risers to ensure that the mold is evenly and completely filled.

After a few minutes the mold cavity is filled and the vacuum equalized. The mold is left closed for the complete curing time, which is typically 45 minutes to four hours. In stage two, the part is ejected and the flash, risers and excess material are removed.

Notes for Designers

QUALITY Surface finish is excellent and will be an exact replica of the surface of the pattern used in the production of the mold.

TYPICAL APPLICATIONS Prototyping applications include automotive parts, live hinges, keyboards and housings for mobile phones, televisions, cameras, MP3 players, sound systems and computers.

COST AND SPEED Tooling costs are generally low, but this is largely dependent on the size and complexity of the pattern. A mold will usually take between half a day to one day to make in silicone and will last 20–30 cycles. Cycle time is good.

MATERIALS PUR is available in water-clear grades and the full colour range. It can be very soft and flexible (shore A range 25–90) or rigid (shore D range).

ENVIRONMENTAL IMPACTS Accurately measuring the material reduces scrap, which cannot be recycled because it is a thermosetting plastic. The process is carried out in a vacuum chamber, so any fumes and gases can be extracted and filtered.

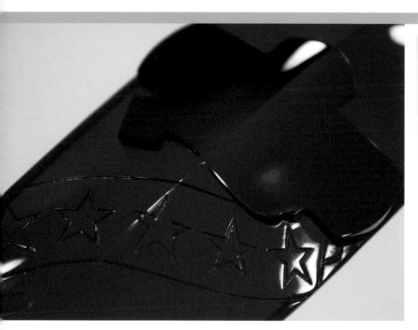

Translucent colour The PUR prototyping material range is designed to mimic injection molded plastics. Ribs, surface textures, snap fits, live hinges and other details that are usually associated with injection molding can be made.

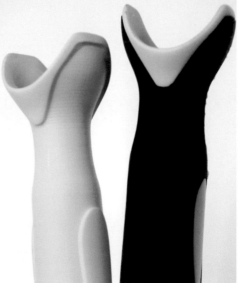

Mimicking multishot injection molding The properties of different plastics can be incorporated as over-moldings. This is similar to multishot injection molding. In this case, rigid yellow acrylonitrile butadiene styrene (ABS) mimic has been over-molded with flexible black thermoplastic elastomer (TPE) mimic.

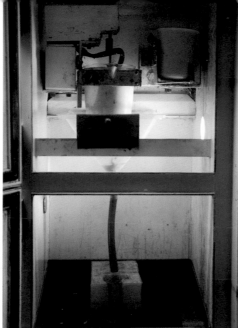

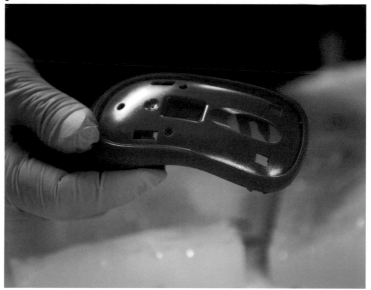

Case Study

Vacuum Casting a Computer Mouse

Featured company CRDM www.crdm.co.uk

Molds are made from silicone rubber. Metal inserts are used to reproduce intricate details and thin wall sections (image **1**).

The casting process takes place under vacuum. The PUR pours into the mold and is unrestricted by air pressure (image **2**). As the mold fills up the PUR emerges from the risers in the mold.

The two halves of the mold are separated and the casting is carefully removed with the risers intact (image **3**). These are detached by hand (image **4**). The finished part is very similar to the original mass-produced injection-molded part, with ribs, perforations, logos and matt surface finish (image **5**).

1

2

3

4

5

Vacuum Forming

Vacuum forming is used to form 3D prototypes and models from thermoplastic sheet materials. A single-sided mold is used, which is relatively inexpensive, and the sheet materials are formed using heat and pressure. A range of thermoplastics from 1 mm to 12 mm (0.04–0.47 in.) thick can be formed with this process.

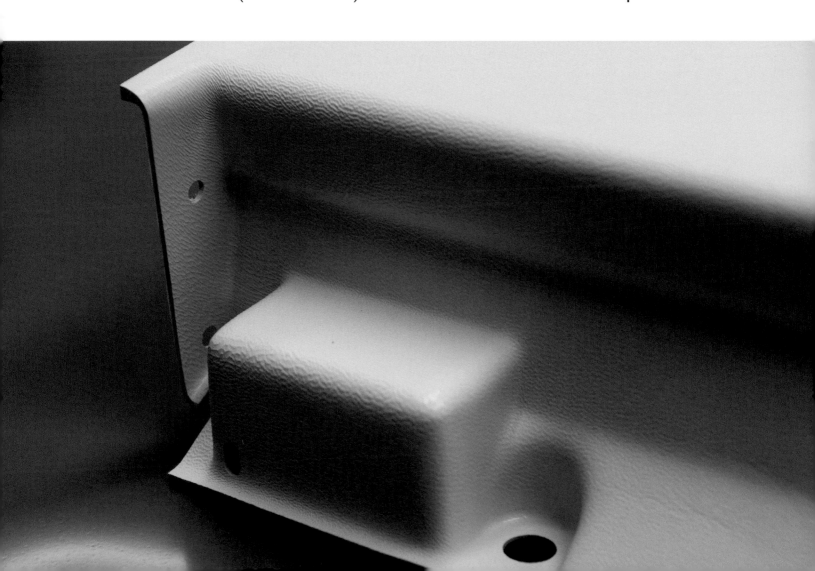

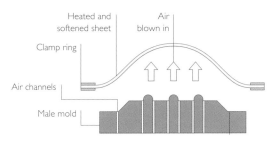

Stage 1: Pre-heated sheet

Heated and softened sheet
Air blown in
Clamp ring
Air channels
Male mold

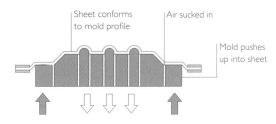

Stage 2: Forming the vacuum

Sheet conforms to mold profile
Air sucked in
Mold pushes up into sheet

What is Vacuum Forming?

A sheet of material is heated to its softening point. This is different for each material. For example, the softening point of polystyrene (PS) is 127–182°C (261–360°F) and polypropylene (PP) is 143–165°C (289–329°F). Certain materials, such as high-impact polystyrene (HIPS), have a larger operating window (that is, the temperature range in which they are formable), which makes them much easier to vacuum form.

In stage one, the softened plastic sheet is blown into a bubble which stretches it in a uniform manner. In stage two, the airflow is reversed and the mold is pushed up into the sheet. A strong vacuum draws the material onto the surface of the mold to form the final shape. To assist the flow of air, channels are drilled into the mold. They are located in recesses and across the surface of the mold to extract the air as efficiently as possible.

Notes for Designers

QUALITY One side of a thermoformed plastic sheet comes into contact with the mold and so will have an inferior finish. However, the reverse side will be smooth and unmarked. Therefore, parts are generally designed so that the side that came into contact with the mold is concealed in application.

TYPICAL APPLICATIONS Typical examples include models and prototypes for packaging, cosmetic trays, drinking cups, briefcases, automotive panels and fridges.

COST AND SPEED Mold-making costs are typically low to moderate, depending on the size, complexity and quantity of parts. Cycle time is reasonably fast. Labour costs are low to moderate depending on the number of manual operations required.

MATERIALS Most thermoplastic materials can be thermoformed. In vacuum forming the mold can be made from metal, wood or resin. Wood and resin are ideal for prototyping and low-volume production.

ENVIRONMENTAL IMPACTS This process is only used to form thermoplastic materials so the majority of scrap can be recycled.

Vacuum forming intricate surface patterns Using microporous aluminium it is possible to vacuum form intricate surface design details such as dimples, or large flat areas, that would otherwise be impractical to vent with air channels. The micropores allow a vacuum to be applied right through the material and onto the hot sheet above. It is generally only utilized for prototypes and short production runs because the micropores quickly become blocked. It is formed by conventional CNC machining (page 42).

1

2

3

Prototypes and low-volume parts are often vacuum formed using polyurethane resin (PUR) molds (image **1**) which are formed by CNC machining (page 42).

The sheet of polyethylene terephthalate modified with glycol (PETG) is prepared by screen printing. The printed details are typically applied to the reverse so the material protects the finish in application. The sheet is loaded face down onto the mold (image **2**) and the forming process takes place.

Once formed, the sheet is removed (image **3**). The finished prototype is compared to an earlier test piece (images **4** and **5**). The graphic details on the finished piece have been aligned with the molding by first printing a greyscale version accompanied by a grid. This shows exactly how the material is stretching during forming and is used to calculate how much compensation is required in the printed graphics so that they align perfectly with the finished 3D form.

4

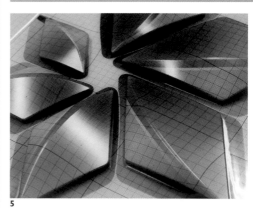

5

Mold Making

Molds are required for casting a wide range of materials. Once a mold has been constructed it can be used to produce multiple identical parts. Molds are an exact reverse of the product being cast and are made by CNC machining or by replicating a 3D pattern using mold-making techniques.

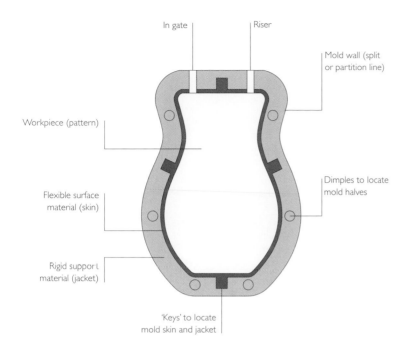

In gate

Riser

Mold wall (split or partition line)

Workpiece (pattern)

Flexible surface material (skin)

Dimples to locate mold halves

Rigid support material (jacket)

'Keys' to locate mold skin and jacket

What is Mold Making?

It is possible to take a mold of almost anything from a suspect's footprint in the mud to giant sculptures. This versatility means mold making is a highly skilled process and can take several days or even weeks to complete.

The major consideration for mold making is ensuring that the pattern and casting can be easily removed. This is achieved by building split lines (also referred to as mold partition lines) into the mold. These make a continuous line around the pattern where the mold halves separate. Complex shapes with undercuts in multiple directions will require multiple split lines. These are carefully worked out so that they do not impact on the physical or aesthetic properties of the final casting.

Notes for Designers

QUALITY The quality of the mold is dependent on the quality of the pattern, skill of the craftsman and required number of castings. Should longer production runs be required then a more durable skin material is used.

TYPICAL APPLICATIONS This technique is used to make works of art, sculptures, statues, prototypes and even car body panels. However, the techniques and materials used vary according to application.

COST AND SPEED The cost of patternmaking depends on the size and complexity of design. It may cost nothing if the mold is being taken from an existing object. Cycle time is moderate to long and labour costs are high.

MATERIALS The chosen mold-making materials depend on the desired casting material and vice versa.

ENVIRONMENTAL IMPACTS Composite molds cannot be recycled easily. Some materials used in the construction of molds are harmful and hazardous when disposed.

Plaster bust with visible split lines Complex profiles, such as a human face, have many undercuts. Silicone is sufficiently elastic to be pulled out of most crevices. However, if a rigid mold is built directly onto the pattern every undercut has to be taken into account. The flash on this bust, evidence of the mold split lines, demonstrates the number of mold parts required to make this casting. Flash can be removed from rigid and semi-rigid materials. However, it is virtually impossible to remove them completely from flexible materials.

1

2

3

4

5

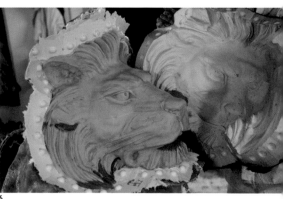

6

Featured company Bronze Age Ltd
www.bronzeage.co.uk

The polyurethane resin (PUR) foam pattern was CNC machined by Bakers Patterns and is finished by hand (images **1** and **2**). It is essential that the surface of the pattern is exactly as the final part will be because the mold will reproduce every detail exactly. In this case a soft silicone 'skin' is applied to the surface of the pattern (image **3**). This is necessary because there are lots of fine details and small undercuts. Making a silicone skin reduces the number of mold parts required. Once the first coat has fully cured the layer of silicone is built up by hand (image **4**) and 'keys' are bonded in place. These will locate the silicone skin into the structural fibreglass jacket that is applied on top (image **5**). The pattern is demolded in stages (image **6**). Afterwards the mold is reassembled and the internal cavity will be an exact replica of the foam pattern. This particular mold is used to cast the wax patterns used in lost wax casting (page 30). The original lions were created by W. W. Wagstaff.

Lost Wax Casting

Also known as investment casting, this is an expensive metal forming process, but the opportunities far outweigh the cost implications for many applications, including prototypes and large-scale artwork. Liquid metals are formed into complex and intricate shapes in this process which uses non-permanent ceramic molds.

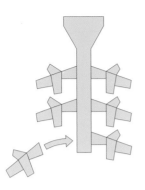

Stage 1: Assemble wax patterns onto tree

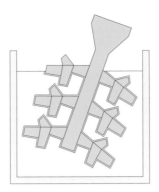

Stage 2: Assembly coated in ceramic slurry

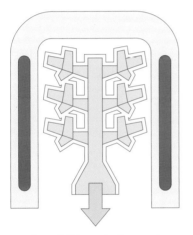

Stage 3: Wax melted out and ceramic mold fired

Stage 4: Metal poured into hot mold

Essential Information

VISUAL QUALITY	●●●●●●○○
SPEED	●●●●●●○○
MOLD AND JIG COST	●●●●●○○
UNIT COST	●●●●●○○○
ENVIRONMENT	●●●●●○○○

Alternative and competing processes include:
- Centrifugal Casting
- Die Casting
- Forging
- Metal Injection Molding
- Rapid Prototyping
- Sand Casting

What is Lost Wax Casting?

In stage one, the expendable patterns are formed and assembled together on a central feed system. In stage two, this is dipped in ceramic slurry and then coated with fine grains of refractory material.

In stage three, the wax patterns and runner system are melted out in a steam autoclave and the ceramic shell is subsequently fired at 1095°C (2003°F).

In stage four, molten metal is poured in. Once the casting has solidified and cooled, it is broken out of the shell mold.

QUALITY Lost wax casting produces high-integrity metal parts with superior metallurgical properties. The surface finish is generally very good and can be improved afterwards.

TYPICAL APPLICATIONS Designers, artists and architects utilize this process to make models, prototypes, sculpture, jewelry, architectural components and furniture.

COST AND SPEED There are moderate tooling costs for the wax-molding process. Cycle time is long and up to several weeks, depending on the size and complexity of the part. A great deal of labour is required and so costs are usually quite high.

MATERIALS Many types of metal can be formed in this way. In low-volume applications the most common materials include copper alloys (brass and bronze), zinc, aluminium and precious metals (such as gold and silver). Rapid prototyping techniques, including selective laser sintering (SLS) (page 105) and stereolithography (SLA) (page 110), can be utilized in the production of the pattern (over which the ceramic shell mold is formed) directly from CAD data.

ENVIRONMENTAL IMPACTS Very little waste is produced that cannot be recycled. However, this is an energy intensive process operating at high temperatures.

Case Study

Lost Wax Casting a Bronze Artwork

Featured company Bronze Age Ltd
www.bronzeage.co.uk

Firstly, artist Stephen Hunter creates a clay sculpture. This is replicated in wax by making a mold around the clay (page 26). As soon as the wax has hardened it is removed from the mold (image **1**).

The wax pattern is slightly larger than the final artwork to allow for shrinkage during casting. It is assembled onto a wax feed system and coated with ceramic slurry (image **2**) followed by fine grains of refractory ceramic powder (image **3**). This is built up in several layers and strengthened with a layer of glass-reinforced plastic (GRP).

The wax pattern is melted out to produce a hollow ceramic mold. Several molds are heated to 750°C (1382°F) and submerged in a chamber filled with sand. Molten bronze is poured in (image **4**) and solidifies.

After only one hour the ceramic shell mold can be broken away, although it will still be very hot and so is often left over night to cool down. The parts of the statue are fettled (flash and runner systems removed) and assembled by TIG welding (page 140) (image **5**).

Finally, the surface goes through a process of articifial patination (page 167), using heat and chemicals to produce a rich, dark and uniform colour (image **6**). The finished statue is installed in London (image **7**).

2

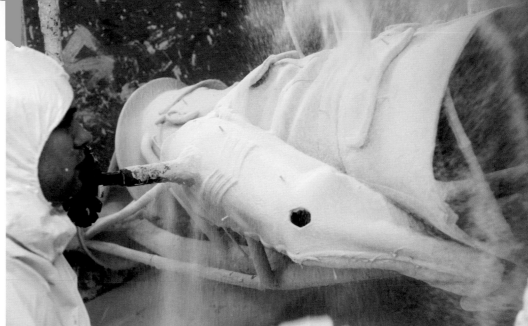

3

4

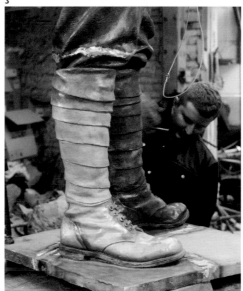

5

6

7

Sand Casting

Sand casting is used to shape molten ferrous metals and non-ferrous alloys in single-use sand molds. It relies on gravity to draw the molten material into the die cavity and so produces rough parts that have to be finished by abrasive blasting, machining or polishing.

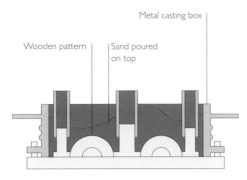

Metal casting box

Wooden pattern

Sand poured
on top

Stage 1: Mold making

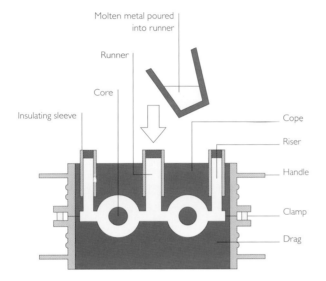

Molten metal poured
into runner

Runner

Core

Insulating sleeve

Cope

Riser

Handle

Clamp

Drag

Stage 2: Sand casting

Essential Information

VISUAL QUALITY	●●●●○○○○○
SPEED	●●●○○○○○
MOLD AND JIG COST	●●●●●○○○
UNIT COST	●●●●●○○○
ENVIRONMENT	●●●●○○○○

Related processes include:
• Dry Sand Casting
• Green Dry Sand Casting

Alternative and competing processes include:
• Centrifugal Casting
• Die Casting
• Forging
• Investment Casting

What is Sand Casting?

The sand casting process is made up of two main stages: mold making and casting. In stage one, the mold is made in two halves, known as the cope and drag.

For dry sand casting the sand has a vinyl ester polymer coating which is room temperature cured. The polymer coating helps to achieve a better surface finish in the cast metal. For green sand casting, the sand is mixed with clay and water until it is sufficiently wet to be rammed into the mold and over the pattern. The clay mix is left to dry so that there is no water left in the mix.

In stage two, the mold is clamped together. The metal is heated up to several hundred degrees above its melting temperature and poured into the runner so it fills the mold completely.

Notes for Designers

QUALITY Sand-cast metal has a distinctive surface finish; so all cast parts are abrasive blasted and polished to achieve a better surface finish. Sand casting relies on gravity to draw the molten material into the mold cavity so there will always be an element of porosity in the cast part.

TYPICAL APPLICATIONS Applications include furniture, lighting, architectural fittings, cylinder heads and engine blocks.

COST AND SPEED Tooling costs are low, the main cost being for patternmaking. Low-cost patterns can be machined in wood or aluminium. Foam patterns are the least expensive of all. Cycle time is moderate but depends on the size and complexity of the part.

MATERIALS This process can be used to cast ferrous metals and non-ferrous alloys. The most commonly sand-cast materials include iron, steel, copper alloys (brass, bronze) and aluminium alloys.

ENVIRONMENTAL IMPACTS Energy requirements for sand casting are quite high because the metal has to be raised to several hundred degrees above its melting temperature. In green sand casting up to 95% of the mold material can be recycled after each use.

Case Study

Sand Casting a Lamp Housing

Featured company Chiltern Casting Company

The mold is prepared by packing sand over a wooden pattern which is removed to reveal the mold cavity (image **1**).

Metal is heated to well above its meting point (image **2**) and poured into the runner where it spreads through the mold and up the risers. When the mold is full an exothermic metal oxide (aluminium oxide in this case) is poured into the runner and risers (image **3**). The powder burns at a very high temperature which keeps the aluminium molten in the top of the mold for longer. This is important to minimize porosity in the top surface of the cast part.

After 15 minutes the casting boxes are separated along the partition line. The sand is broken away from around the metal casting (image **4**). The parts are cut free from the runner system and prepared for finishing (image **5**).

2

3

4

5

Panel Beating

Panel beating is controlled forming of sheet metal using hand tools, sandbags, profiled rollers and jigs. Smooth curves and undulating shapes are possible: combined with metal welding and other sheet-metal forming processes, panel beating by a skilled operator is capable of producing almost any shape.

Metal workpiece

Wooden or nylon mallet

Bag of sand or metal shot

Dishing into a sandbag

Metal workpiece

Nylon or metal chaser

Engineer's hammer

Epoxy or steel jig

Jig chasing

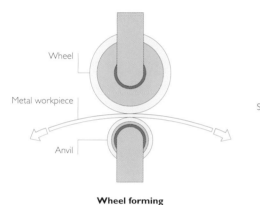

Wheel

Metal workpiece

Anvil

Wheel forming

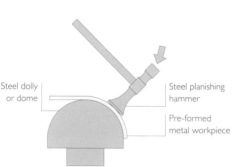

Steel dolly or dome

Steel planishing hammer

Pre-formed metal workpiece

Planishing

Essential Information

VISUAL QUALITY	●●●●●●○○
SPEED	●●○○○○○○
MOLD AND JIG COST	●●●●●○○○
UNIT COST	●●●●●●○○
ENVIRONMENT	●●●○○○○○

Related processes include:
- Dishing
- Jig Chasing
- Planishing
- Wheel Forming

Alternative and competing processes include:
- Metal Press Forming
- Metal Spinning
- Superplastic Forming

What is Panel Beating?

Bags of sand or metal shot are still used for certain applications. Dishing is rapid, but it is the least accurate and controllable of the panel beating techniques.

Jig chasing (also known as hammerforming) is the process of stretching and compressing sheet metal to conform to the shape of a jig. The jig is either 'soft' and made of epoxy, or 'hard' and made of steel.

Wheel forming is also known as English Wheeling. The metal workpiece is passed back and forth between a wheel and an anvil. The wheel is flat faced and the anvil has a profile (crown). The role of the anvil is to stretch the sheet progressively with each overlapping pass.

Planishing is a finishing operation and is essentially smoothing over the surface with repeated and overlapping hammer blows.

Notes for Designers

QUALITY A skilled operator can achieve a superior 'A-class' finish with a combination of planishing and polishing. These techniques are used to finish stainless steel brightwork for Bentley production cars because the requirements of the surface finish are so high.

TYPICAL APPLICATIONS It is used in the automotive, aerospace and furniture industries for prototyping, pre-production and low-volume production runs.

COST AND SPEED Mold-making costs are low to moderate depending on the size and complexity. Cycle time is moderate: it is possible to construct the chassis and bodywork of a car from 3D CAD drawings in approximately six weeks. This is a labour intensive process and the level of skill required is very high.

MATERIALS Most ferrous and non-ferrous metals can be shaped in this way.

ENVIRONMENTAL IMPACTS Panel beating is an efficient use of materials and energy. There is no scrap produced in the forming operations, although there may be scrap produced in the preparation and subsequent finishing operations.

Dishing into a sand bag Dishing into a sand bag is now largely confined to prototyping.

Austin-Healey 3000 This Austin-Healey 3000 was given a completely new body by CPP (Manufacturing) Ltd.
  Almost any shape can be produced in metal by panel beating. Large and small radius curves are produced with similar ease by a skilled operator. Sheets can be embossed, beaded or flanged to improve their rigidity without increasing their weight.

1

Panel Beating the Spyker

Featured company CPP (Manufacturing) Ltd
www.cpp-uk.com

This case study illustrates the production of the Spyker C8 Spyder in aluminium (image **1**).

A sheet is cut to size and passed back and forth between the wheel forming rolls in overlapping strokes (image **2**). Each pass stretches the metal slightly and so forms a two-directional bow in the sheet.

When the correct curvature is approximately achieved in the sheet it is transferred onto the jig chaser. It is clamped onto the surface of the epoxy jig and gradually stretched and compressed to conform to the shape (images **3** and **4**).

The panels are shaped individually and then brought together on a jig. They are TIG welded (page 140) to form strong and seamless joints (image **5**).

2

3

4

5

CNC Machining

CNC machining encompasses a range of shaping processes and is used to manufacture precise and high-quality models, products and artwork. It is used to shape metal, plastic, wood, stone, composite and other materials into large undulating shapes and technical products with intricate details.

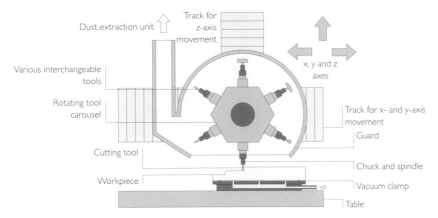

Track for z-axis movement

Dust extraction unit

Various interchangeable tools

Rotating tool carousel

Cutting tool

Workpiece

x, y and z axes

Track for x- and y-axis movement

Guard

Chuck and spindle

Vacuum clamp

Table

Three-axis CNC with tool carousel

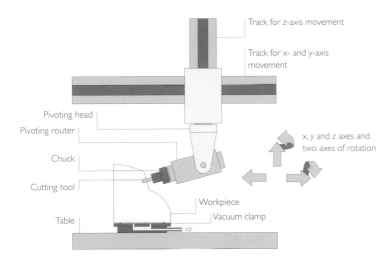

Track for z-axis movement

Track for x- and y-axis movement

Pivoting head

Pivoting router

Chuck

Cutting tool

Table

x, y and z axes and two axes of rotation

Workpiece

Vacuum clamp

Five-axis with interchangeable tools

Essential Information

VISUAL QUALITY	●●●●●●○○
SPEED	●●○○○○○○
MOLD AND JIG COST	○●●●●●○○
UNIT COST	●●●●●●●○
ENVIRONMENT	●●●○○○○○

Related processes include:
• CNC Lathe Turning
• CNC Milling
• CNC Routing

Alternative and competing processes include:
• CNC Turret Punching
• EDM (Electrical Discharge Machining)
• Laser Cutting
• Rapid Prototyping
• Reaction Injection Molding
• Vacuum Casting
• Veneer Laminating

What is CNC Machining?

Among the many different types of CNC machinery, CNC milling machines and CNC routers are essentially the same. CNC lathes, on the other hand, operate differently because the workpiece is spun rather than the tool. The woodworking and metalworking industries will probably use different names for similar tools and operations – the names and practices can be traced back to when these materials were hand worked using material-specific tools and equipment.

CNC machinery has x- and y-axis tracks (horizontal) and a z-axis track (vertical).

Many different tools are used in the cutting process, including cutters (side or face), slot drills (cutting action along the shaft as well as the tip for slotting and profiling), conical, profile, dovetail and flute drills, and ball nose cutters (with a dome head, which is ideal for 3D curved surfaces and hollowing out). By contrast, CNC lathes use single-point cutters because the workpiece is spinning.

Notes for Designers

QUALITY CNC machining produces high-quality parts with close tolerances. Cutting traces can be reduced or eliminated by, for example, grinding, sanding and polishing (page 170) the part.

TYPICAL APPLICATIONS CNC machining is a cost-effective method for producing models, prototypes, patterns and tools for other processes such as lost wax casting (page 30) and reaction injection molding (page 14), film props and theatre sets.

COST AND SPEED Tooling costs are minimal and are limited to jigs and other clamping equipment. Cycle time is rapid once the machines are set up.

MATERIALS Almost any material can be CNC machined, including plastic, metal, wood, glass, ceramic and composites.

ENVIRONMENTAL IMPACTS This is a reductive process, so generates waste in operation. Modern CNC systems have very sophisticated dust extraction which collects all the waste for recycling or incinerating for heat and energy use. Dust that is generated can be hazardous, especially because certain material dusts become volatile when combined.

A sculpture by Zaha Hadid Architects This sculpture in Hong Kong, designed by Zaha Hadid Architects, has a CNC machined foam core that is coated with glass-reinforced plastic by A M Structures. This is a cost-effective method used in the manufacture of striking limited edition furniture, sculptures, canopies and reception desks, for example. Almost any shape and finish can be achieved using this technique.

Model making foam Virtually any material can be shaped by CNC machining. These samples of polyurethane resin (PUR) are used for model making, patterns and tools. The density ranges from very light open-cell foams to rigid blocks.

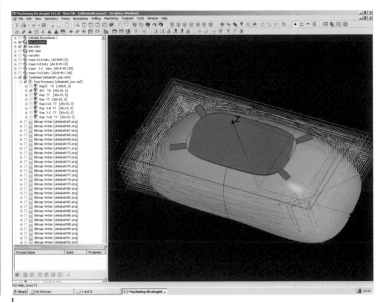

1

CNC Machining a Polystyrene Prototype

Featured company Bakers Patterns Ltd
www.polystyrenemodels.co.uk

CAD files are typically STL format. A cutting path is calculated and the correct cutting tools are selected to ensure efficient processing and to achieve the desired finish (image **1**). For complex parts and intricate details this can take many hours. In this case a block of the expanded polystyrene (EPS) is loaded onto the machine bed and 'roughed out' using a high-speed cutting tool. The basic shape of the product can now be seen (image **2**). The cutting tool is changed to produce a smooth and uniform surface finish (image **3**).

Once the underside has been completed, the model is turned over, centred onto the machine bed, and the internal profile is 'roughed out' (images **4** and **5**). The finished part will have a smooth finish inside and out (image **6**). This can be further improved by sanding and coating with resin. The waste materials are collected, compacted and recycled.

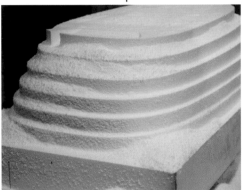

2

3

4

5

6

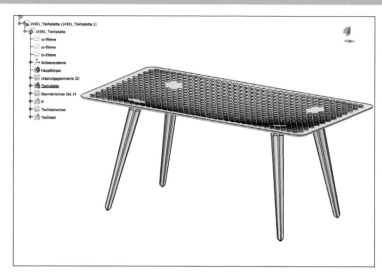

Colander Table CAD file CNC machining is a computer-aided manufacturing (CAM) process. In other words, all CNC machining operations are computer guided. Every detail of the CAD file is translated into a cutting path and the selection of the appropriate cutting tool, which can take many hours for complex geometries. Therefore, it is essential that the CAD file is exactly as the designer envisions the prototype or product because it will be reproduced exactly by the CAM process. It also means that each product can be different without affecting manufacturing time or quality.

Case Study

CNC Machining the Colander Table

Featured company Daniel Rohr www.daniel-rohr.com

The Colander Table was created by German designer Daniel Rohr in 2005 (image **1**). The table top is CNC machined from solid aluminium and weighs 408 kg (899 lb). The four legs are CNC lathed separately.

It takes around four weeks to manufacture the table from start to finish, including the polishing. Firstly, an aluminium ingot is loaded onto the machine bed and clamped in place (image **2**). A tool for rough scrubbing is used to machine the convex top profile in overlapping strokes (image **3**). The rough shape is refined with a finer cutting tool and overlapping strokes, and the holes are machined (image **4**). To help keep the temperature down large amounts of lubricant and coolant are sprayed onto the cutting tool and aluminium.

The top profile is finished and the partly machined ingot is turned over. The backside of the tabletop is machined (image **5**). Once a smooth finish has been achieved the final shape is carefully mirror polished (image **6**) and the legs are fixed in place.

I

2

3

4

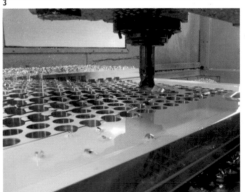

5

6

CNC Turret Punching

Circular, square and profiled holes are cut from sheet materials by a computer-guided steel punch. These techniques are shearing processes and are combined with sheet-metal forming and joining technologies to produce a range of geometries for prototyping and low-volume production.

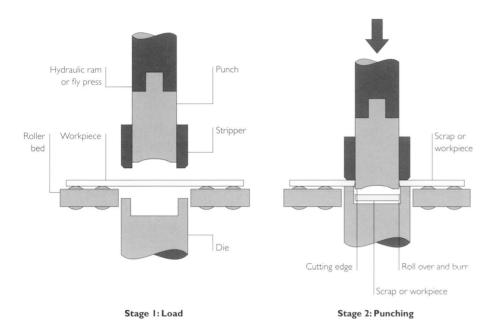

Hydraulic ram
or fly press

Punch

Roller
bed

Workpiece

Stripper

Die

Stage 1: Load

Scrap or
workpiece

Cutting edge

Roll over and burr

Scrap or workpiece

Stage 2: Punching

Essential Information

VISUAL QUALITY	●●●●●○○○○
SPEED	●●●●●○○○○
MOLD AND JIG COST	●●○○○○○○○
UNIT COST	●●●●●○○○○
ENVIRONMENT	●●●●○○○○○

Related processes include:

• Blanking
• Punching

Alternative and competing processes include:

• CNC Machining
• Laser Cutting
• Water Jet Cutting

What is CNC Turret Punching?

The operation is the same whether it is carried out on a turret punch, punch press or flypress. It is possible to punch a single hole, multiple holes simultaneously, or many holes with the same punch.

In stage one, the workpiece is loaded onto the roller bed. In stage two, the stripper and die clamp the workpiece. The hardened punch stamps through it, causing the metal to fracture between the circumferences of the punch and die.

Once cut, the punch retracts and the stripper ensures that the metal comes free. Either the punched material or the surrounding material is scrap, depending on whether it is a punching or blanking operation. In both cases the scrap is collected and recycled.

Notes for Designers

QUALITY This is a precise cutting technique. The shearing action forms 'roll over' on the cut edge and fractures the edges of the material. This results in burrs which are sharp and have to be removed by grinding and polishing (page 170).

TYPICAL APPLICATIONS Some typical products include prototyping and low-volume production of consumer electronic and appliance enclosures, filters, washers, hinges, general metalwork and automotive body parts.

COST AND SPEED Standard and small tools are inexpensive. Cycle time is rapid. Between one and 100 punches can be made every minute. Labour costs are moderate to high.

MATERIALS Almost all metals can be processed in this way. It is most commonly used to cut carbon steel, stainless steel and aluminium and copper alloys and other materials, including leather, textiles, plastic, paper and card.

ENVIRONMENTAL IMPACTS Parts can be nested very efficiently on a sheet to minimize scrap. Any scrap is collected and separated for recycling, so there is very little wasted material.

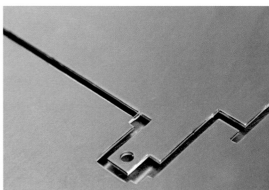

Tabs (above) Parts that are cut out have to be tied into the sheet of material by tabs. These ensure that the part does not move about during production, but they are sufficiently fragile to allow the parts to be broken out by hand. The burr left behind will have to be polished off, unless it is concealed in a recess.

Metal enclosures (left) CNC turret punching is often combined with press braking (page 124) to form metal enclosures like these examples that feature fixing points, air vents and access windows.

Case Study

CNC Turret Punching an Aluminium Blank

Featured company Cove Industries
www.cove-industries.co.uk

The turret punch is loaded with a series of matching punches and dies (image **1**). The orange surround on the punch is the stripper which makes sure the punch can retract from the workpiece. The die set is loaded into the turret and their location programmed into the computer-guiding software (image **2**).

Each cycle produces a small piece of scrap which is siphoned into a collection basket for recycling (image **3**). In a blanking operation these pieces of metal are the workpiece.

The cut-out part is inspected and cleaned up (image **4**). In this case, the metal blanks are formed into enclosures by press braking (page 124).

2

3

4

EDM

The two techniques are die sink EDM (Electrical Discharge Machining), also known as spark erosion, and wire EDM, also known as wire erosion. These precise processes are used to machine metal and apply a surface texture simultaneously. They can be used to produce geometries that are not possible with conventional CNC machining.

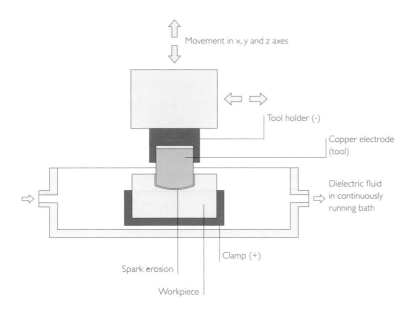

Movement in x, y and z axes

Tool holder (-)

Copper electrode (tool)

Dielectric fluid in continuously running bath

Clamp (+)

Spark erosion

Workpiece

What is Die Sink EDM?

Die sink EDM takes place with the electrode (tool) and workpiece submerged in light oil, similar to paraffin. This dielectric fluid is continuously running to maintain the working temperature and flush out vaporized material.

The copper electrode (tool) and metal workpiece are brought within close proximity which initiates the spark erosion process. High-voltage sparks leap from the electrode to the workpiece and vaporize the surface of the metal. The electrical discharges jump between the closest points on the electrode and the workpiece, forming continuous and even surface material removal.

Notes for Designers

QUALITY The quality of EDM parts is so high they can be used to manufacture tooling for injection molding without any finishing operations. Parts can be manufactured accurately to five microns (0.00019 in.).

TYPICAL APPLICATIONS It has been widely adopted by the toolmaking industry for injection molding, metal casting and forging. It is also used for model making, prototyping and low-volume production of typically no more than 10 parts.

COST AND SPEED Wire EDM does not require tooling. The tooling for die sink EDM is relatively inexpensive, but for precision parts new tools are required for each operation. Cycle time is moderate and comparable to CNC machining, but depends on surface finish.

MATERIALS Metals including stainless steel, tool steel, aluminium, titanium, brass and copper are commonly shaped in this way.

ENVIRONMENTAL IMPACTS This process requires a great deal of energy to vaporize the metal workpiece and fumes are given off during operation, which can be hazardous.

Injection molding The die cavities in these injection-molding tools were formed by die sink EDM. Die sink EDM can be used to produce internal geometries on parts that are not possible with conventional machining. This is because a negative copper electrode can be machined into shapes that are not suitable for cavities. The negative electrode (tool) is reproduced in the workpiece, creating sharp corners and complex features that would otherwise be impractical.

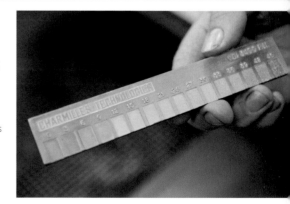

Scale of surface finish Surface finish is measured on the Association of German Engineers' VDI scale. The VDI scale is comparable with roughness average (Ra) 0.32–18 microns (0.000013–0.00071 in.).

Case Study

Die Sink EDM

Featured company Hymid Multi-Shot Ltd
www.hymid.co.uk

In this case study, a cavity is being formed in high carbon steel. It is not practical to machine complex and intricate cavities into hard metals like this, other than by using EDM.

The copper alloy tool (electrode) is inserted into a tool holder and aligned with the metal workpiece (image **1**). Sparks and fumes are given off during rough cutting (image **2**). There are several thousand sparks per second. Each spark vaporizes a small piece of surface material. The resulting surface finish is very rough (image **3**).

To achieve a much finer surface finish the later stages of machining are significantly slower (image **4**).

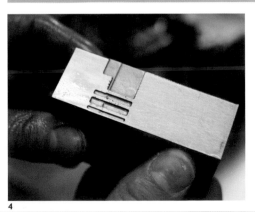

What is Wire EDM?

In this process the wire electrode, which is usually copper or brass, is fed between the supply spool and take-up spool. The wire is held under tension to cut straight lines. The guide heads move in tandem along x and y axes to produce profiles and independently along x and y axes to produce tapers. It is charged with a high voltage, which discharges as the wire progresses through the workpiece. Similar to die sink EDM, the spark occurs in the smallest gap between the metals.

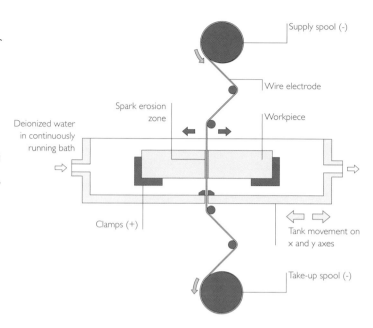

Supply spool (-)

Wire electrode

Spark erosion zone

Deionized water in continuously running bath

Workpiece

Clamps (+)

Tank movement on x and y axes

Take-up spool (-)

Die sink EDM tool shaped by wire EDM Wire EDM can be used in much the same way as hot-wire cutting polymer foam, although it is considerably more precise and much slower. This very small die sink EDM tool was cut out using the wire EDM process. It is too small and intricate to be produced by conventional CNC machining.

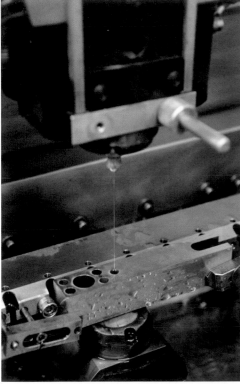

1

2

3

4

Featured company Hymid Multi-Shot Ltd
www.hymid.co.uk

The partly machined high carbon steel workpiece is loaded into a jig and clamped in place. A small hole is drilled in preparation: this is where the wire electrode is fed through (image **1**).

Both the workpiece and wire electrode are submerged in deionized water which acts as an insulator (image **2**). Once submerged, the cutting process begins (image **3**).

On the finished article you can just make out the kerf where the wire has cut from the pre-drilled hole to the cutting profile (image **4**): on the left are the parts prior to cutting, in the centre is the finished workpiece and on the right the material that has been removed.

Electroforming

Electroforming is the same as electroplating, but it is carried out on non-conductive materials. It is made possible by covering the workpiece with a layer of silver particles which form a surface onto which metal is deposited. The object being electroformed can be used as a mold, or encapsulated.

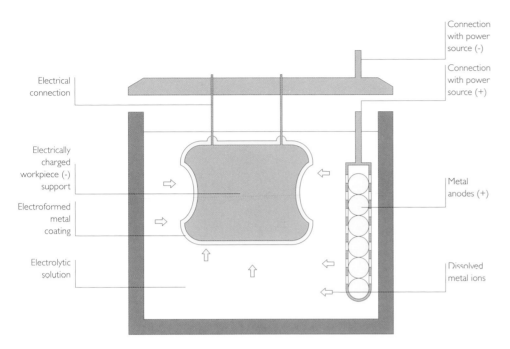

Electrical
connection

Electrically
charged
workpiece (-)
support

Electroformed
metal
coating

Electrolytic
solution

Connection
with power
source (-)

Connection
with power
source (+)

Metal
anodes (+)

Dissolved
metal ions

What is Electroforming?

The silver coating on the surface of the mandrel is connected to a DC power supply. This causes suspended metal ions in the electrolytic solution to bond with it and build up a layer of pure metal.

As the thickness of electroplating builds up on the surface of the workpiece, the ionic content of the electrolytic solution is constantly replenished by dissolution of the metal anodes, which are suspended in the electrolyte in a perforated conductive basket.

Wall thickness is precisely controlled and can range from 5 microns (0.00019 in.) up to 25 mm (0.98 in.) in thickness.

Notes for Designers

QUALITY The surface finish is an exact replica of the surface of the mandrel in reverse. Therefore, quality of finish is determined by the original product.

TYPICAL APPLICATIONS Due to its exceptional accuracy, it is popular for biomedical equipment, microsieves and razor foils. Decorative applications include film props, sculptures and jewelry.

COST AND SPEED Mold-making costs are inexpensive for low volumes compared to metal press forming tools. The cycle times are moderate and much faster than hand engraving. For high volumes it is a slow and expensive means of production. Labour costs are moderate to high.

MATERIALS Almost any material, such as wood, ceramics and plastic, can be electroformed or encapsulated with metal. Any object or material that cannot be used as a mandrel can be reproduced in silicone. Electroformed metals include nickel, copper, silver and gold.

ENVIRONMENTAL IMPACTS Electroforming is an additive rather than reductive process. In other words, only the required amount of material is used. However, plenty of hazardous chemicals are required for electroforming.

Metal encapsulation This follows the same basic electroforming process, except that the workpiece is covered entirely by a uniform conductive silver layer and so cannot be reused. In this case, a wooden carving has been encapsulated with a thin layer of gold.

1

2

Featured company BJS Company www.bjsco.com

The silicone rubber mold, taken from a hand carving, is coated with pure silver powder (image **1**). The surface coating is connected to a DC power supply and the mold is submerged in the electroforming tank (image **2**), where the electrolytic solution is supplied with fresh ions by pure copper anodes (image **3**). The copper is deposited on the surface of the mold to form a uniform wall thickness.

After 48 hours this part is fully formed (image **4**) and has a wall thickness of 1 mm (0.04 in.).

3

4

Ceramic Wheel Throwing

Ceramic products that are symmetrical around an axis of rotation can be made on a potter's wheel. The style, shape and function of each piece can be as varied as the potter who creates it, and each studio and craftsman adapts and develops their own techniques.

Rotating clay pot

Bat

Potter's wheel

Essential Information

VISUAL QUALITY	●●●●○○○○
SPEED	●●○○○○○○
MOLD AND JIG COST	●○○○○○○○
UNIT COST	●●●●●●○○
ENVIRONMENT	●●●●●○○○

Alternative and competing processes include:
- Ceramic Press Forming
- Ceramic Slip Casting
- Jiggering and Jolleying

What is Ceramic Wheel Throwing?

A predetermined quantity of clay is thrown onto a 'bat' which is then placed on the potter's wheel. The ball of clay is centred on the spinning wheel which is generally powered by an electric motor.

While the wheel turns the potter gradually draws the clay vertically upwards to create a cylinder with an even wall thickness. Clay throwing must always start in this way, to ensure even wall thickness and distribution of stress, even though the shape can subsequently be manipulated into a variety of geometries.

Ceramic products made in this way are fired twice. Firstly, they are biscuit fired to remove all the moisture and prepare the ceramic for glazing. Finally, they are glaze-fired in a kiln at temperatures in the region of 1200–1400°C (2192–2552°F).

Notes for Designers

QUALITY Since it is handmade, thrown pottery tends to have variable quality that depends on the skill of the potter and on the material itself. Stoneware and porcelain have much better mechanical properties than earthenware and so can be thrown with much thinner wall sections.

TYPICAL APPLICATIONS Wheel throwing is used to produce a wide range of gardenware, kitchenware and tableware.

COST AND SPEED There are no mold-making costs. Cycle time is moderate, but depends on the complexity and size of the part. The firing time can be quite long and is determined by whether the parts are biscuit fired and then glaze-fired, or are once-fired only. Labour costs are moderate to high.

MATERIALS Earthenware, stoneware and porcelain can be thrown on a potter's wheel. Porcelain is the most difficult material to throw and earthenware the easiest because it is more robust and forgiving.

ENVIRONMENTAL IMPACTS There are no harmful by-products from this pottery-forming process. However, the firing process is energy intensive and therefore the kiln is fully loaded for each firing cycle. Once-firing reduces the energy used.

Applying colour Glaze is used to apply colour, decorative effects and to make earthenware watertight. Ceramic glazes are typically made up of silica combined with a metal oxide pigment, such as copper, cobalt or iron. Firing in a kiln at high temperature causes the silica and metal oxide to melt and form a smooth glass coating over the surface of the ceramic.

All types of shapes By the very nature of this process, all parts will be rotationally symmetrical. To create asymmetric shapes other skills, such as handwork, carving and pressing, are combined with clay throwing. Knobs, feet, spouts and other embellishments are added after throwing.

1

2

3

4

Case Study

Wheel Throwing Porcelain Cups and Saucers

Featured company Rachel Dormor Ceramics
www.racheldormorceramics.com

A predetermined measure of porcelain is worked by hand into the correct consistency for throwing (image **1**). Pummelling the clay is essential and levels the density of the material and makes it supple for shaping.

The lump of porcelain is rotated and centred by hand and the potter makes a doughnut shape (image **2**). The porcelain is drawn up into a cylinder with an even wall thickness and then shaped into the cup geometry (image **3**).

The cups are left to air-dry for about an hour before the handles are bonded on with liquid porcelain. Once the parts are sufficiently leather-hard they are ready for the first firing (image **4**).

Jiggering and Jolleying

These ceramic flatware techniques for manufacturing multiple replica parts are used in the production of kitchen and tableware, including pots, cups, bowls, dishes and plates. Although they can all be automated, these techniques are often still carried out by hand.

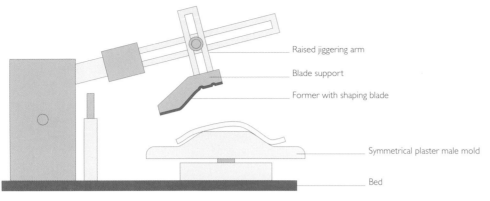

Raised jiggering arm

Blade support

Former with shaping blade

Symmetrical plaster male mold

Bed

Stage 1: Open mold, loading and unloading

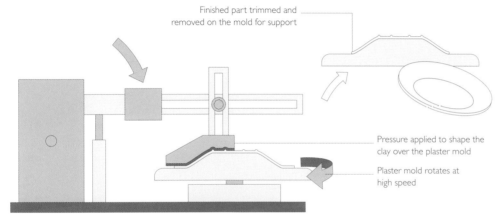

Finished part trimmed and
removed on the mold for support

Pressure applied to shape the
clay over the plaster mold

Plaster mold rotates at
high speed

Stage 2: Closed mold

Essential Information

VISUAL QUALITY	●●●●●●○○○○
SPEED	●●●●●○○○○○
MOLD AND JIG COST	●●○○○○○○○○
UNIT COST	●●●●○○○○○○
ENVIRONMENT	●●●●○○○○○○

Alternative and competing processes include:
- Ceramic Press Molding
- Ceramic Slip Casting
- Ceramic Wheel Throwing

What is Jiggering?

The process of jiggering is known as jolleying if the mold is in contact with the outside surface of the part rather than the inside surface. A plaster 'male' mold is used in the jiggering process and a 'female' one for the jolleying process.

In stage one, a charge of clay is loaded onto a mold spinning at high speed. In stage two, the jiggering arm is brought down onto the clay. A profiled former with a shaping blade forces the clay to take the shape of the mold. The process is very rapid and takes less than a minute.

QUALITY A very high-quality surface finish can be achieved with both jiggering and jolleying. The pottery materials used in press molding are brittle and porous, so the surface is often made vitreous by glazing, which provides a watertight seal.

TYPICAL APPLICATIONS Applications include flatware (such as plates, bowls, cups and saucers, dishes and other kitchen and tableware vessels), sinks and basins, jewelry and tiles.

COST AND SPEED Mold-making costs for jiggering and jolleying are relatively low. Cycle time is rapid and a skilled operator can spin around one part per minute. Labour costs are moderate.

MATERIALS Clay materials, including earthenware, stoneware and porcelain, can be pressed.

ENVIRONMENTAL IMPACTS In all pressing operations scrap is produced at the 'green' stage and so can be directly recycled. There are no harmful by-products from these pottery-forming processes. However, the firing process is energy intensive, so therefore the kiln is fully loaded for each firing cycle.

Molds used for jiggering (above) Each design requires a different mold and if large volumes are required then multiple molds are used because the formed parts need to be left on the mold to air-dry before biscuit firing.

Surface details reproduced with jiggering (left)
Jiggering is used to produce large volumes of identical parts. It is possible to reproduce intricate surface details and patterns on the surface that comes into contact with the plaster mold. Details can be embossed and pierced onto the outside surface (or inside surface in the case of jolleying) while the clay is 'green'.

1

2

3

4

5

6

Case Study

Jiggering a Plate

Featured company Hartley Greens & Co.
(Leeds Pottery) www.hartleygreens.com

Firstly, a 'pancake' of clay is spun with an even wall thickness (image **1**). This is transferred onto the spinning plaster mold and the jiggering arm is brought down onto the clay to form the outside surface (image **2**).

Once the final shape is complete and the edges have been trimmed, the mold and clay are removed from the metal carrier intact (image **3**). The clay part is left on the mold until it is sufficiently 'green' to be removed.

The part is biscuit fired to remove all remaining moisture. This takes place in a kiln over eight hours. The temperature of the parts is raised and then soaked at 1125°C (2057°F) for one hour before cooling slowly. Following the first firing, the plates are removed from the molds and the edges are finished (images **4** and **5**).

All remaining surface decoration is carried out, such as glazing and hand painting, before the biscuitware is glaze-fired (image **6**).

Kiln Forming Glass

This glass-forming technique is also referred to simply as 'slumping' or 'drape forming' and is the process of heating glass in a kiln until it becomes soft enough to drape into a mold or over a form under its own weight. A vacuum or jig is used for added precision. It is utilized in the production of both engineered and handcrafted objects.

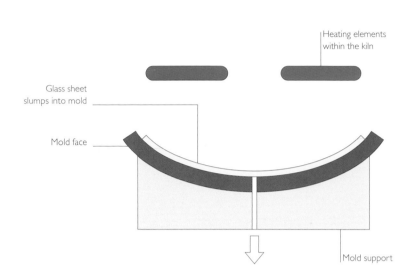

Glass sheet
slumps into mold

Mold face

Heating elements
within the kiln

Mold support

Essential Information

VISUAL QUALITY	●●●●●●●○○
SPEED	●●●○○○○○○
MOLD AND JIG COST	●●●○○○○○
UNIT COST	●●●●●●●○○
ENVIRONMENT	●●●●○○○○

Related processes include:
- Fusing
- Laminating

Alternative and competing processes include:
- CNC Machining
- Lampworking
- Studio Glassblowing

What is Kiln Forming Glass?

The mold is manufactured in a refractory or suitable material that is capable of withstanding the high temperatures reached in the kiln. Sheet glass is slumped into the mold, or over a form. The surface finish on the mold face will determine the finish on the glass that makes contact with it. The other side of the glass sheet will remain smooth and glossy.

The temperature of the kiln, typically 650–850°C (1202–1562°F), and the time taken to heat and anneal the glass depends on the type of glass being formed.

The exact temperature and time is critical if the glass being slumped is textured, or if multiple colours are being fused together during forming, for example.

It may be necessary to carry out several firing operations if coloured glasses are being fused together, or if a complex bent profile in multiple directions is to be achieved. A vacuum can be introduced into the forming cycle to increase precision, reduce cycle time and form sharper bends, and pivoting jigs can be used to reduce stretching in thick sections.

Notes for Designers

QUALITY The mold surface finish will impact on the final surface finish and texture of the glass, as will the temperature and forming time. The quality of artistic and fused work will depend on the skill of the craftsman.

TYPICAL APPLICATIONS Applications are wide ranging and include car windows, curved architectural façades, glass screens, furniture, lighting, sculpture and tableware.

COST AND SPEED Mold-making costs are low to moderate and depend on the size and complexity of the part. Cycle time is long (larger kilns hold more parts, but take longer to heat and cool). Labour costs are moderate to high and depend on the type of work being undertaken.

MATERIALS All types of glass can be kiln formed, including soda-lime glass (float glass), lead alkali glass, borosilicate glass and high-performance glasses. It is generally not possible to combine two different types of glass because their coefficients of expansion will be different and so will cause the part to shatter when it cools.

ENVIRONMENTAL IMPACTS There is very little scrap produced during kiln forming, however, this is a high-temperature process and so is energy intensive.

Kiln forming In these pieces Cathryn Shilling has combined fusing with forming to create elegant and beautiful works of art. First of all, the coloured glass, which must be compatible, is fused together by kiln firing. Secondly, the fused glass is kiln formed into the desired shape. Multi-coloured and decorative pieces typically require two or three firing cycles to achieve all the desired effects.

1

2

3

4

5

6

Case Study

Kiln Forming Glass Light Diffusers

Featured company Instrument Glasses
www.instrumentglasses.com

The molds are supported by a refractory ceramic base (image **1**), which ensures that the thin mold does not deform during forming. The metal molds are loaded onto the bases (image **2**) and the pre-cut sheets of soda-lime glass are placed on top (image **3**). The textured side is laid face down into the mold.

The glass discs are formed in the kiln at 700°C (1292°F). At this point a delicate balance is reached whereby the glass conforms to the shape of the mold and the texture remains intact. The parts are removed from the kiln and mold (images **4** and **5**). The whole process takes around 24 hours.

The finished parts are stacked and packed ready for shipping (image **6**).

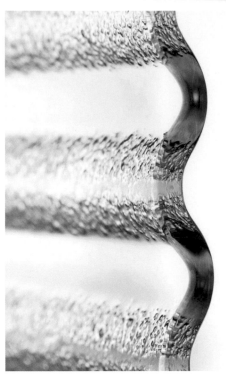

Coloured glass sheet It is possible to form coloured glass and combine many layers, or fragments, for decorative applications such as screens and furniture. Glass sheets ranging from 25–130 mm (0.98–5.12 in.) are suitable for kiln forming.

Textured glass The surface of the mold will determine the surface finish on the side of the glass that it comes into contact with. Many different textures can be applied and further enhanced with abrasive blasting, colouring and mirroring (page 80).

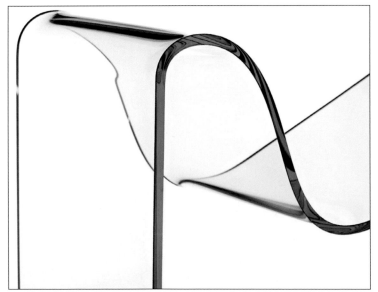

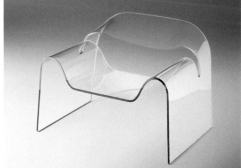

Ghost Chair It is possible to achieve multiple bends in different directions. This is not easy but is made possible by carefully designed and constructed molds, intricately controlled firing cycles and skilful craftsmanship. This is demonstrated in the Ghost Chair, which was designed by Cini Boeri in 1987 for FIAM Italia.

Drape Forming the Genio Coffee Table

Featured company FIAM Italia www.fiamitalia.it

Designed by Massimo Iosa Ghini in 1991 and manufactured by FIAM Italia, the Genio Coffee Table (image **1**) is formed in either 10 or 12 mm (0.4 or 0.47 in.) thick soda-lime glass.

This is a precision process: it is essential that the glass remains perfectly flat up to the start of the bend. To achieve this, the glass is draped over a carefully constructed mold (image **2**). Heatproof fabric made from woven glass fibre protects the surface of the glass against the steel mold.

The sheet glass is placed onto the mold (image **3**) and restrained by two pivoting jigs (image **4**). These prevent the heavy glass from over stretching during forming.

1

2

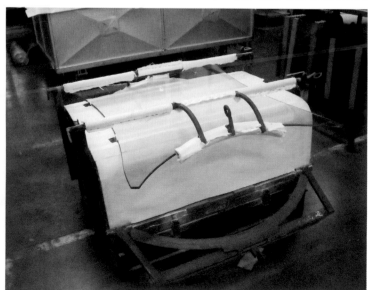

3

4

Studio Glassblowing

Both decorative and functional, hollow and open-ended vessels can be blown in glass. The process involves blowing a bubble of air inside a mass of molten glass, which is either blown and shaped by hand with formers, or blown and shaped in a mold.

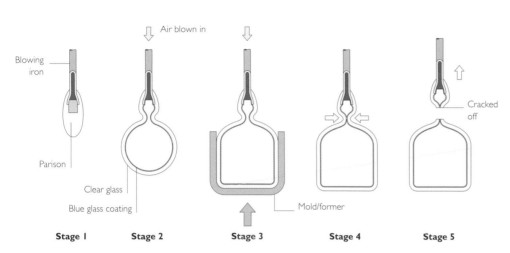

Blowing
iron

Air blown in

Parison

Clear glass

Blue glass coating

Mold/former

Cracked
off

Stage 1   **Stage 2**   **Stage 3**   **Stage 4**   **Stage 5**

Essential Information

VISUAL QUALITY	●●●●●●○○
SPEED	●●○○○○○○
MOLD AND JIG COST	●●●●●●○○
UNIT COST	●●●●●○○○
ENVIRONMENT	●●●●○○○○

Related processes include:
• Glassblowing
• Mold Glassblowing

Alternative and competing processes include:
• CNC Machining
• Kiln Forming Glass
• Lampworking

What is Studio Glassblowing?

In stage one, the nose of the blowing iron is pre-heated to over 600°C (1112°F) and loaded with a piece of coloured glass (if colour is being used). This is dipped into a crucible of molten glass.

In stage two, air is blown in and it is intermittently inserted into the glory hole to maintain its temperature above 600°C (1112°F). This is a gas-fired chamber which is used to keep the glass at a working temperature.

In stage three, profiled formers or molds may be used to shape the glass accurately. In stage four, pucellas (sprung metal tongs) are used to reduce the diameter of the neck of the glass vessel so it can be cracked off and transferred to the kiln for annealing.

Notes for Designers

QUALITY Glass is a material that has a high perceived value because it combines decorative qualities with great inherent strength. The structure is weakened only by surface imperfections and impurities in the raw material.

TYPICAL APPLICATIONS A variety of vessels, containers and bottles, including tableware and cookware.

COST AND SPEED Mold-making costs are low to non-existent for studio glassblowing because equipment costs are low. Studio glass is a slow process that requires great skill and experience.

MATERIALS Soda-lime glass is the most commonly used. Light shades, tableware, cut glass, crystal glass and decorative objects are typically made from lead alkali glass. Borosilicate glass is used for laboratory equipment, high-temperature lighting applications and cookware.

ENVIRONMENTAL IMPACTS Glass is a long-lasting material and a great deal of recycled materials is used in production. However, glassblowing is energy intensive and so there have been many developments in recent years to reduce energy consumption.

Etched glass vessel Because the studio glass processes are free from the limitations of mass production, there are many ways that glassmakers can manipulate the shape and surface of the glass to create exciting effects such as sophisticated etched patterns, cracked surfaces and encapsulated air bubbles.

This glass vessel, which was blown into a mold by the studio team at The National Glass Centre (see opposite), was designed by Peter Furlonger in 2005.

Case Study

Studio Glassblowing with a Mold

Featured company National Glass Centre
www.nationalglasscentre.com

Hot glass is gathered onto the end of the blowing iron and is shaped into a parison using cherry wood formers and paper which have been soaked in water (image **1**). This process is repeated several times.

The hot glass parison is then laid into a dish of blue powdered glass (image **2**). Layers of colour are built up in this way to create added 'depth' in the abrasive blasted finish, which is applied later (see opposite).

The glass parison is repeatedly blown, heated in a glory hole (image **3**) and shaped until it is a suitable size with adequate wall thickness for molding.

The mold is pre-heated to around 600°C (1112°F) in a small kiln. The blown glass parison is placed in the mold, rotated and blown simultaneously, forcing it against the relatively cooler mold walls, which starts to harden the glass (image **4**). Soon afterwards, the finished vessel is cut from the blown-glass form (image **5**).

1

2

3

4

5

Wide Stone from the Ariel range by Peter Layton

Peter Layton is renowned for his use of colour to produce dynamic and engaging works in glass.

The molten glass can be marked and decorated in many ways. Coloured glass and silver foils can be rolled onto the surface and then encapsulated into the part with another layer of clear glass. Coloured threads of glass can be trailed over the surface of the parison from a separate gather. These trails can be dragged over the surface of the glass to create patterns, in a technique known as 'feathering'.

Case Study

Studio Glassblowing with Coloured Effects

Featured company London Glassblowing
www.londonglassblowing.co.uk

The white glass is heated to around 800°C (1472°F) and a gob of molten red and clear glass is gathered onto a punty and allowed to run over the white glass (image **1**). Overlaying coloured glass builds up strata that add visual depth to the final piece.

A thin stream of molten blue glass is trailed across the surface of the overlaid gob in a spiral (image **2**).

To enhance the pattern even further, the molten mass of glass is wound around a punty, which coils the glass and transfers the spiralling pattern from longitudinal to helix (image **3**).

The glass parison, which has now been transferred to a blowing iron, is dipped into the crucible in the furnace and a 'gather' of clear glass forms a coating over the pattern (image **4**), which is worked in a wood block or former (image **5**).

Once blowing is complete, the parison is transferred onto a punty and the end is opened out to form a bowl (image **6**). After the bowl is finished he cracks it off the end of the punty and it is transferred to an annealing kiln for controlled cooling over a period of 36 hours (image **7**).

1

2

3

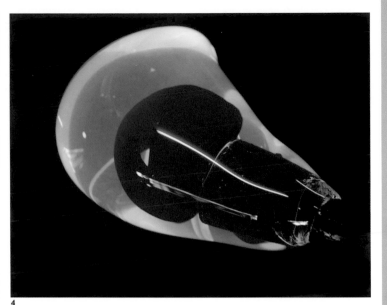

4

5

6

7

Lampworking

Glass is formed into hollow shapes and vessels by lampworking, also known as 'flameworking', using a combination of intense heat and manipulation by a skilled lampworker. Products range from jewelry to complex scientific laboratory equipment. Lampworking is carried out as either benchwork or lathework.

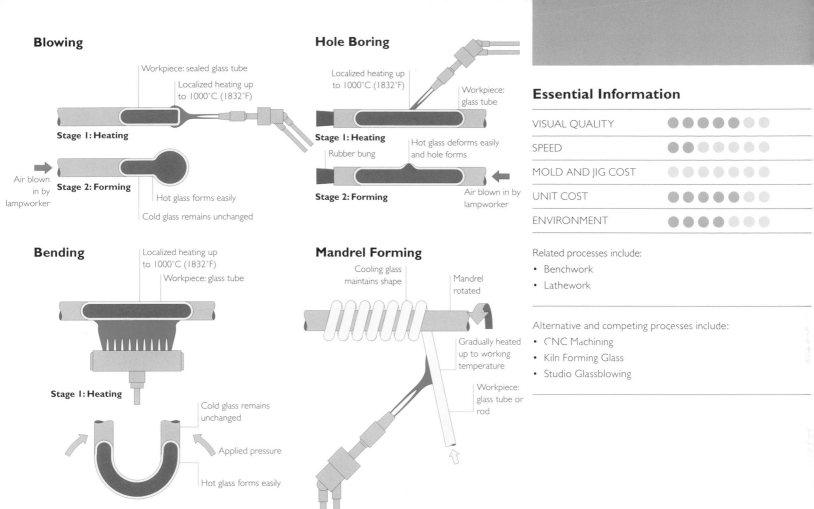

Blowing

Workpiece: sealed glass tube

Localized heating up to 1000°C (1832°F)

Stage 1: Heating

Air blown in by lampworker

Stage 2: Forming

Hot glass forms easily

Cold glass remains unchanged

Hole Boring

Localized heating up to 1000°C (1832°F)

Workpiece: glass tube

Stage 1: Heating

Rubber bung

Hot glass deforms easily and hole forms

Stage 2: Forming

Air blown in by lampworker

Bending

Localized heating up to 1000°C (1832°F)

Workpiece: glass tube

Stage 1: Heating

Cold glass remains unchanged

Applied pressure

Hot glass forms easily

Stage 2: Forming

Mandrel Forming

Cooling glass maintains shape

Mandrel rotated

Gradually heated up to working temperature

Workpiece: glass tube or rod

Essential Information

VISUAL QUALITY	●●●●●●●○○
SPEED	●●●○○○○○○
MOLD AND JIG COST	●○○○○○○○○
UNIT COST	●●●●●○○○
ENVIRONMENT	●●●●○○○○

Related processes include:
- Benchwork
- Lathework

Alternative and competing processes include:
- CNC Machining
- Kiln Forming Glass
- Studio Glassblowing

What is Benchworking?

A mixture of natural gas and oxygen are burnt to generate the heat required for lampworking. The working temperature is 800–900°C (1472–1652°F) for borosilicate glass, at which stage it has the consistency of softened chewing gum.

The tools used are similar to those for glassblowing: various formers shape the workpiece and 'marver' the hot glass.

Everything made in this way has to be annealed in a kiln. For borosilicate glass the temperature of the kiln is brought up to approximately 570°C (1058°F) and then maintained for 20 minutes, after which it is slowly cooled down to room temperature. This process is essential to relieve built-up stress within the glass. Some very large pieces, such as artworks, can take several weeks to anneal, although this is unusual.

QUALITY The quality of the part is largely dependent on the skill of the lampworker. Even tiny stresses in the part will cause the glass to shatter or crack. Annealing the parts after lampworking ensures they are stabilized and stress free.

TYPICAL APPLICATIONS It is used to make specialized scientific apparatus, precision glassware, jewelry, lighting and sculpture.

COST AND SPEED There are usually no mold-making costs, while cycle time is moderate, but depends on the size and complexity of the part. The annealing process is usually run overnight, for up to 16 hours, but can take a lot longer depending on material thickness. Labour costs are high due to the level of skill required.

MATERIALS All types of glass can be formed by lampworking. The two main types are borosilicate and soda-lime glass.

ENVIRONMENTAL IMPACTS Glass scrap can be directly recycled. However, a great deal of heat is required to bring the glass up to working temperature and for annealing.

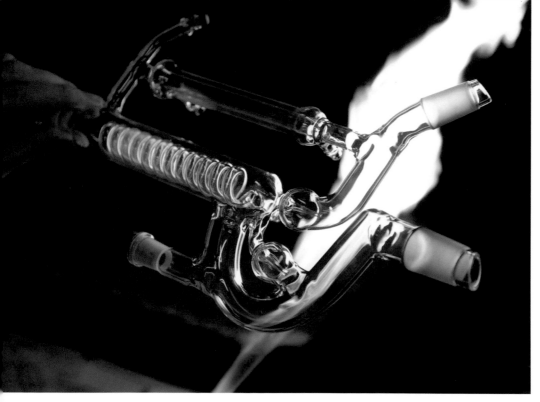

A finished total condensation stillhead The size, geometry and complexity of a part are limited only by the imagination of the designer. Lampworking is equally suitable for precise functional objects and decorative fluid artefacts.

A total condensation stillhead (variable take-off pattern) is a piece of scientific apparatus used for a specific distillation process. The case study opposite demonstrates some of the techniques used to produce it. Like any other lampworking, the stillhead starts as a series of glass tubes.

Case Study

Benchworking a Total Condensation Stillhead

Featured company Dixon Glass
www.dixonglass.co.uk

The first part of the stillhead is brought up to working temperature and starts to glow cherry red. It is blown by hand (image **1**). At each stage of the process the accuracy of the parts is checked against the drawing (image **2**). Tungsten tweezers are used to pull glass across a surface or to bore holes.

A glass spiral is created using a graphite-coated mandrel. The glass must reach its optimum temperature just before it is formed over the mandrel (image **3**).

A U-bend is then worked in another tube section, using a much larger area of heat. Spreading the heat out makes sure that the U-bend will be even across a large diameter. As soon as the glass is up to temperature it is carefully bent by hand (image **4**) to form the final bend (image **5**).

1

2

3

4

5

What is Latheworking?

Similar to benchworking, the glass is heated up to working temperature, which for borosilicate glass is 800–900°C (1472–1652°F).

Throughout the forming process the glass tube is spun at around 60 rpm. Various formers are used to shape the glass as it turns. Benchworking techniques can be used on the glass when it is stationary to create features that are not rotationally symmetrical.

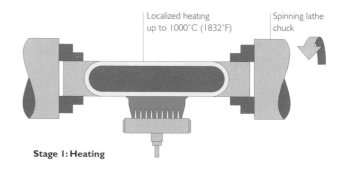

Localized heating up to 1000°C (1832°F)

Spinning lathe chuck

Stage 1: Heating

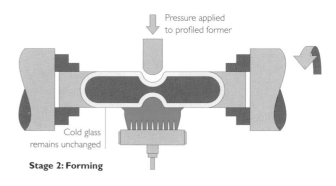

Pressure applied to profiled former

Cold glass remains unchanged

Stage 2: Forming

A triple-walled reaction vessel Latheworking can be used to form large and small parts very accurately. The only requirement is that they are rotationally symmetrical up to a certain point (the final part may have additions that are not symmetrical, created using benchworking techniques) as with the triple-walled reaction vessel.

Case Study

Latheworking the Triple-walled Reaction Vessel

Featured company Dixon Glass
www.dixonglass.co.uk

The operation starts with a large tube section being heated gently to raise its temperature, while being spun slowly. The neck of the innermost jacket is formed with intense localized heat and a profiled carbon bar (image **1**). The two halves of the tube section are separated, and a hole is opened out using a profiled carbon bar (image **2**).

The first two reaction vessel jackets are assembled off the lathe (image **3**).

A hole is opened out and the two jackets are fused together using a carbon bar (image **4**). The process is repeated for the third and final jacket.

Once all three jackets of the reaction vessel are brought together then an extension tube is fused onto the bottom, using the heat of the gas torch in much the same way as benchworking (image **5**).

1

2

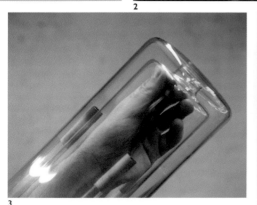

3

4

5

Veneer Laminating

There is nothing new about the process of bonding two or more layers of material together to form a laminate. However, as a result of developing stronger, more water-resistant and temperature durable adhesives, lighter and more reliable structures can now be engineered in laminated wood.

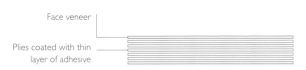

Face veneer

Plies coated with thin
layer of adhesive

Stage 1: Veneer preparation

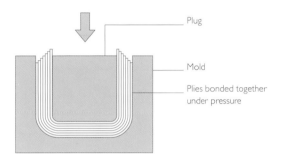

Plug

Mold

Plies bonded together
under pressure

Stage 2: Cold pressing

Essential Information

VISUAL QUALITY	●●●●●●○
SPEED	●●○○○○○
MOLD AND JIG COST	●●●●○○○
UNIT COST	●●●●●○○
ENVIRONMENT	●●○○○○○

Related processes include:
• Bag Pressing
• Cold Pressing

Alternative and competing processes include:
• CNC Machining
• Composite Laminating
• Steam Bending
• Wood Joinery

What is Veneer Laminating?

This process involves a split mold, or mold and plug. The lay-up is symmetrical, with a core made up of an uneven number of plies and face veneers of a similar material and equal thickness. This is essential to ensure that the part does not warp.

In stage one, adhesive is applied to the face of each veneer as it is laid on top of the last. In stage two, the plies are grouped together and pressed into the mold by a plug.

The adhesives are cured by low-voltage heating, radiant heating, radio frequency (RF), or at room temperature.

QUALITY The quality is high, although the parts often require finishing operations and sanding. The integral quality of the parts is determined by the grade of timber and strength and distribution of the adhesive.

TYPICAL APPLICATIONS Products include furniture and architectural products, for use both indoors and outdoors.

COST AND SPEED Mold-making costs are low to moderate. Although cycle time can be long, it depends on the adhesive curing system. RF adhesive curing is generally between two and 15 minutes, for example. Labour costs are moderate to high for manual operations.

MATERIALS Any timber that is cut into veneers can be laminated. The most flexible timbers include birch, beech, ash, oak and walnut.

ENVIRONMENTAL IMPACTS These processes generally have a low impact, especially if the wood is sourced locally and from renewable sources. The various laminating processes require different amounts of energy.

Cold pressing the Isokon Long Chair Marcel Breuer designed the Isokon Long Chair in 1936.

The shape of the arms is too complex to be laminated in a single split mold. Therefore, the mold comprises several parts that are progressively clamped together to form the final profile.

The minimum internal radius of a bend is determined by the thickness of the individual veneers, rather than by the number of veneers or thickness of the build.

1

2

Cold Pressing the Flight Stool

Featured company Isokon Plus
www.isokonplus.com

The Flight Stool was designed by
BarberOsgerby in 1998 (image **1**).
It is laminated in a split mold and the
adhesive curing is accelerated with radio
frequency (RF).

The birch core veneers and walnut face
veneers are prepared with adhesive (image
2). They are loaded into the metal-faced
mold (image **3**). A copper coil is inserted
to connect the metal-faced mold halves
(image **4**). RF generation is activated, which
raises the temperature of the adhesive to
approximately 70°C (158°F) by exciting
the molecules. This accelerates the curing
process so that the part can be removed
from the mold within 10 minutes.

The part is demolded (image **5**). The
Flight Stool is trimmed, sanded and painted.

3

5

What is Bag Pressing?

This process, which uses a vacuum to force the part onto a single-sided mold, reduces costs and increases flexibility. However, only shallow geometries can be formed in this way.

Similar to cold pressing, in stage one, adhesive is applied to the face of each veneer as it is laid on top of the last. In stage two, the plies are grouped together and drawn onto a single-sided mold by a vacuum. Adhesive curing is accelerated by radiant heating.

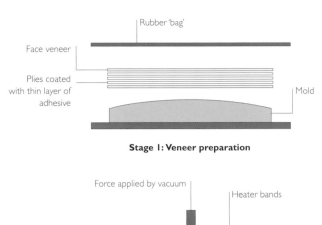

Stage 1: Veneer preparation

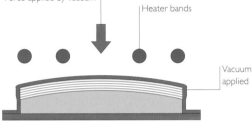

Stage 2: Bag pressing

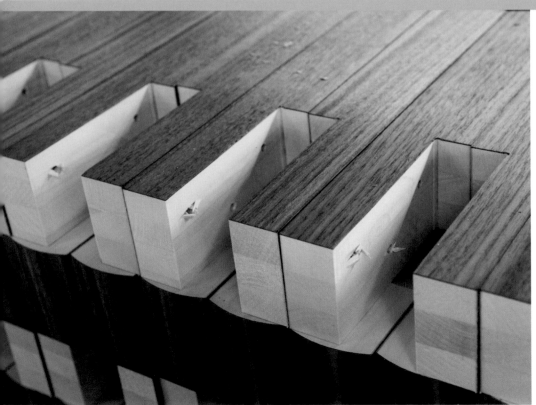

Veneer laminating onto solid timber Expensive, rare or exotic timbers are often sliced into veneers and laminated onto a solid timber substrate for decorative purposes. This reduces the cost considerably. In this case, a relatively expensive walnut veneer has been laminated onto a much less expensive timber substrate. Bag pressing is used to overcome the subtle concave profile on the outside surface of the design. The solid timber provides the necessary mechanical strength.

1

2

3

Case Study

Bag Pressing the Donkey3

Featured company Isokon Plus
www.isokonplus.com

The Donkey3 (image **1**) was designed by Shin and Tomoko Azumi in 2003 and is a development of the original Isokon Penguin Donkey designed by Egon Riss in 1939.

The birch veneers are prepared and a film of adhesive is applied to each surface. The veneers are then laid on a single-sided mold (image **2**). The rubber seal is drawn over the parts (image **3**) and a vacuum forces the lamination to take the shape of the mold (image **4**). A heater is introduced to raise the temperature on the mold to 60°C (140°F) and decrease cycle time. After 20 minutes the adhesive is fully cured and the parts can be removed from the mold (image **5**).

4

5

Composite Laminating

Strong fibres and rigid plastics can be amalgamated to form ultra lightweight and robust products using composite laminating. A range of material combinations is used to produce parts that are suitable for the demands of high-performance applications such as racing cars, aeroplanes and sailing boats.

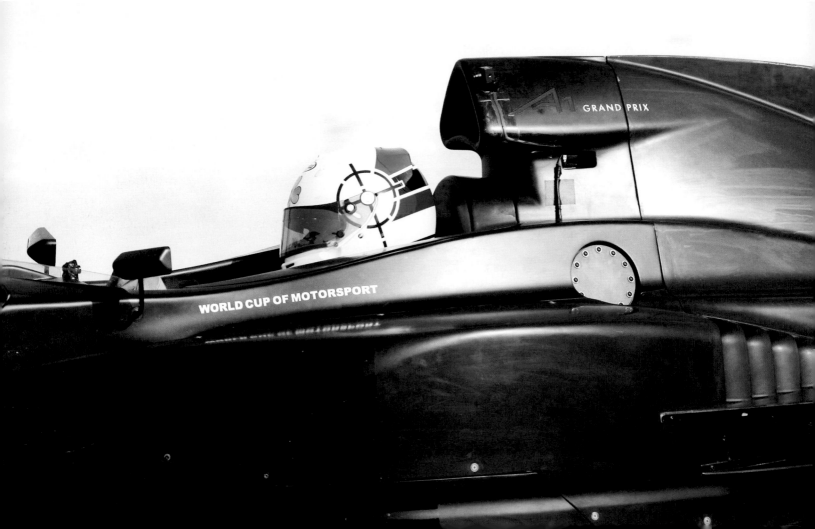

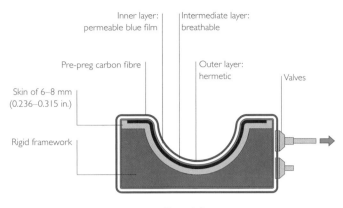

Inner layer:
permeable blue film

Intermediate layer:
breathable

Pre-preg carbon fibre

Outer layer:
hermetic

Valves

Skin of 6–8 mm
(0.236–0.315 in.)

Rigid framework

Stage 1: Lay-up

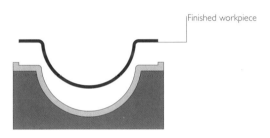

Finished workpiece

Stage 2: Demolding

What is Pre-preg Carbon Fibre Lay-up?

In stage one, layers of carbon fibre pre-impregnated with epoxy resin (pre-preg) are laid up in the mold. Then the whole mold is covered with three layers of material. The first is a blue film, which is permeable, while the intermediate layer is a breathable membrane. These two are sealed in with a hermetic film and a vacuum is applied. These layers ensure that an even vacuum can be applied to the whole surface area.

The pre-preg lay-up is placed into an autoclave, which is raised to a pressure of 4.14 bar (60 psi) and a temperature of 120°C (248°F) for two hours. In stage two, the finished part is demolded.

Notes for Designers

QUALITY The mechanical properties of the product are determined by the combination of materials and lay-up method. Only the side in contact with the mold can be gloss.

TYPICAL APPLICATIONS Applications are becoming more widespread and include racing cars, boat hulls and the structural framework in aeroplanes and furniture.

COST AND SPEED Mold-making costs are moderate to high because the process can be labour intensive and requires a high degree of skill. Cycle time varies: a small part might take an hour or so, whereas a large complex one may require as much as 150 hours.

MATERIALS Fibre reinforcement materials include glass, carbon and aramid. Laminating thermosetting resins include polyester, vinylester and epoxy.

ENVIRONMENTAL IMPACTS Harmful chemicals are used in the production and it is not possible to recycle any of the offcuts or scrap material directly. Material developments are reducing the environmental impact of the process. For example, hemp is being researched as an alternative to glass fibre.

Lola B05/30 Formula 3 racing car In total, this car is made up of hundreds of carbon fibre components. Construction is closely tied into design and engineering. High-performance products have to be engineered to take the maximum load, yet be as lightweight as possible. It is the role of a carbon fibre engineer to push carbon fibre to its limits.

Carbon fibre Carbon fibre has higher heat resistance, tensile strength and durability than glass fibre. When combined with a precise amount of thermosetting plastic it has an exceptional strength to weight ratio, which is superior to steel. Carbon fibre twill (pictured) is the most common weave.

Carbon fibre front wing The lightweight front wing of the Lola B05/30 Formula 3 racing car is made up of carbon fibre laminated over a foam core (pictured).
    Core materials are used to increase the depth of the parts and thus increase torsional strength and bending stiffness, without a big increase in weight.

Pre-Preg Carbon Fibre Lay-up

Featured company Lola Cars International
www.lolacars.com

This case study demonstrates the production of the roll hoop training edge on an Intersport Lola racing car (image **1**).

The carbon fibre is cut out using a CNC plotter. It comes coated with plastic film on either side. The yellow side is peeled off just prior to lay-up (image **2**). Each carbon fibre pattern is aligned on the part and rubbed down (image **3**).

When all the layers are in place the assembly is sealed within a vacuum bag (image **4**) and the air is drawn out. The vacuumed mold is then placed in an autoclave (image **5**), which cures the resin with heat and pressure.

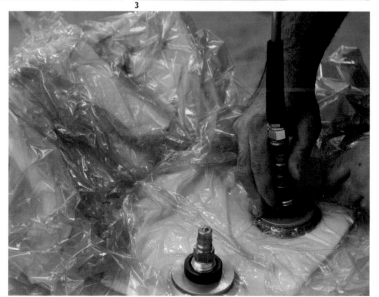

What is Wet Lay-up?

In stage one, a gel coat is applied to the surface of the single-sided mold. Gel coats are anaerobic thermosetting resins; in other words, they cure when not in the presence of oxygen, which is ideal for the mold face.

Mats of woven fibre reinforcement are laid onto the gel coat and then thermosetting resin is painted or sprayed onto it. It is important to achieve the right balance of resin to fibre reinforcement. Rollers are used to remove porosity.

In stage two, the part is demolded. The side in contact with the surface of the mold is typically the outside surface of the finished product.

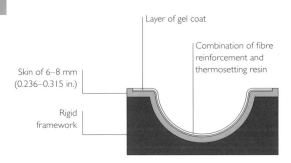

Layer of gel coat

Combination of fibre reinforcement and thermosetting resin

Skin of 6–8 mm (0.236–0.315 in.)

Rigid framework

Stage 1: Lay-up

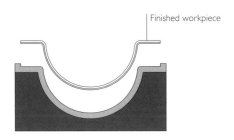

Finished workpiece

Stage 2: Demolding

Glass fibre furniture (above) Glass fibre is a general purpose laminating material often used in hand lay-up processes. It is heat resistant, durable, has good tensile strength, is relatively inexpensive and can be used for a range of applications, such as these lightweight stacking stools, designed by Rob Thompson. Non-woven materials are the least expensive and known as chop strand mat.

Woven glass fibre (right) Weaves include plain (known as 0–90) (pictured), twill and specialist.

1

Composite Laminating the Ribbon Chair

Featured company Radcor www.radcor.co.uk

The Ribbon chair was designed by Ansel Thompson in 2002 (image **1**). It is constructed with vinylester, glass and aramid reinforcement, and a polyurethane foam core.

The parts of the mold are assembled (image **2**) and a wax release agent is applied. A gel coat is painted onto the mold surface and allowed to air-dry before the masking tape is removed (image **3**).

Sheets of fibre are laid onto the gel coat (image **4**), and vinylester is applied to the back. The layers are gradually built up.

The mold is clamped shut (image **5**). After about 45 minutes the vinylester is fully cured and the mold is separated. The chair is removed from the mold, cleaned and de-flashed, in preparation for spray painting and polishing (image **6**).

2

3

4

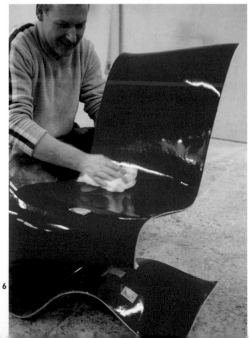

5

6

Filament Winding

Layers of carbon fibre monofilaments, coated in epoxy resin, are wound onto a shaped mandrel. The mandrel is either removed and reused, or permanently encapsulated by the carbon fibres. Filament winding is used for applications that demand the high-performance characteristics of fibre-reinforced composites.

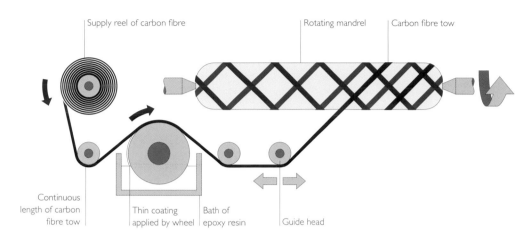

Supply reel of carbon fibre

Rotating mandrel

Carbon fibre tow

Continuous
length of carbon
fibre tow

Thin coating
applied by wheel

Bath of
epoxy resin

Guide head

Essential Information

VISUAL QUALITY	●●●●●●○○
SPEED	●●●●○○○○
MOLD AND JIG COST	●●●●○○○○
UNIT COST	●●●●●○○○
ENVIRONMENT	●●●●●●○○

Related processes include:
• Encapsulation (Bottle Winding)
• Mandrel Winding

Alternative and competing processes include:
• 3D Thermal Laminating
• Composite Laminating
• Compression Molding
• Plastic Extrusion

What is Filament Winding?

The carbon fibre tow is applied to the rotating mandrel by a guide head. The head moves up and down along the mandrel as it rotates and guides the filament to form the geodesic overlapping pattern.

This is the wet lay-up process; the continuous length of fibre reinforcement is coated with an epoxy resin by a wheel rotating in a bath of the resin.

A complete circuit is made when the guide head has travelled from one end of the mandrel to the other and back to the starting point. The speed of the head relative to the speed of mandrel rotation will determine the angle of the fibre.

Notes for Designers

QUALITY Stiffness is determined by the thickness of lay-up and tube diameter. Winding is computer guided, making it precise to 100 microns (0.0039 in.). The angle of application will determine whether the layer is providing longitudinal, torsional (twist) or circumferential (hoop) strength.

TYPICAL APPLICATIONS Application examples include blades for wind turbines and helicopters, pressure vessels, deep-sea submersibles, suspension systems and structural framework for aerospace applications.

COST AND SPEED Mandrel costs are low to moderate. Encapsulating the mandrel will increase the cost. Winding cycle time is 20 to 120 minutes for small parts. Curing time is typically four to eight hours. Labour costs are moderate.

MATERIALS Types of fibre reinforcement include glass, carbon and aramid. Resins are typically thermosetting and include polyester, vinylester, epoxy and phenolics.

ENVIRONMENTAL IMPACTS Thermosetting materials cannot be recycled, so scrap and offcuts have to be disposed of. However, new thermoplastic systems are being developed which will reduce the environmental impacts of the process.

Encapsulating a metal liner with carbon fibre composite Three-dimensional hollow products that are closed at both ends can be made by winding the carbon fibre over a hollow liner, which becomes integral to the final product. This technique is known as bottle winding and is used to produce pressure vessels, housing and suspension systems.

Other than shape, the benefits of winding over a liner include forming a water-, air- and gas-tight skin.

The gloss surface finish is a gel coat resistant to heat or chemical attack.

1

2

3

4

5

Filament Winding a Racing Propshaft

Featured company Crompton Technology Group
www.ctgltd.co.uk

The angle of application ranges from 90° to almost 0° (image **1**). In this case the carbon fibre is sealed onto the mandrel with plastic tape (image **2**). The tape squeezes excess epoxy resin from the carbon and ensures a high-quality, smooth finish.

The filament wound assemblies are placed into an oven, which cures the resin at up to 200°C (392°F) for four hours. Small droplets of resin form on the surface of the tape during curing (image **3**). This is removed when the tape is peeled off. The wound carbon tube is taken off the mandrel (image **4**).

These components are used in cars to deliver power from the engine to the wheels. Traditionally they are made from metal, but when carbon composite is used it is possible to make weight savings of almost three quarters (image **5**).

Rapid Prototyping

These layer-building processes have revolutionized the design industry. There is no tooling and parts are built directly from CAD data. A range of technologies exist for building with powders and liquids. In fact, many materials, such as polymers, ceramics, wax, metals and even paper, can be formed by rapid prototyping.

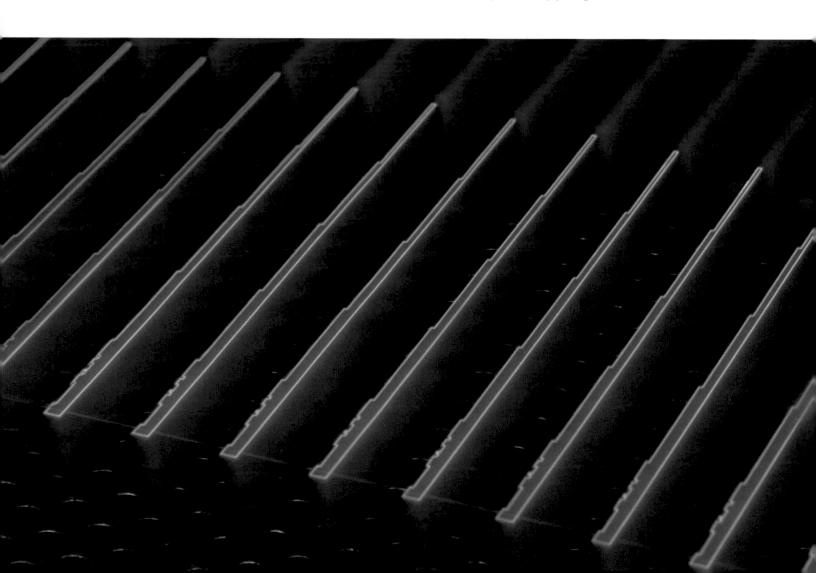

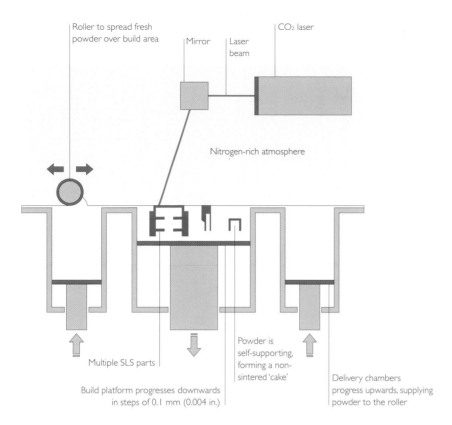

Roller to spread fresh
powder over build area

Mirror

Laser
beam

CO₂ laser

Nitrogen-rich atmosphere

Multiple SLS parts

Build platform progresses downwards
in steps of 0.1 mm (0.004 in.)

Powder is
self-supporting,
forming a non-
sintered 'cake'

Delivery chambers
progress upwards, supplying
powder to the roller

Essential Information

VISUAL QUALITY	●●●●●●○○
SPEED	●●○○○○○○
MOLD AND JIG COST	●○○○○○○○
UNIT COST	●●●●●●○○
ENVIRONMENT	●●○○○○○○

Related processes include:
• Direct Metal Laser Sintering (DMLS)
• Selective Laser Sintering (SLS)
• Stereolithography (SLA)

Alternative and competing processes include:
• CNC Machining
• Lost Wax Casting
• Mold Making
• Vacuum Casting

What is Selective Laser Sintering (SLS)?

In this layer-additive manufacturing process, a CO_2 laser fuses fine nylon powder in 0.1 mm (0.004 in.) layers, directed by a computer-guided mirror. The build platform progresses downwards in layer thickness steps. The delivery chambers alternately rise to provide the roller with a fresh charge of powder to spread accurately over the surface of the build area. The whole process takes place in an inert nitrogen atmosphere at less than 1% oxygen to stop the nylon oxidizing when heated by the laser beam.

The temperature inside the building chamber is maintained at 170°C (338°F), just below the melting point of the polymer powder, so that as soon as the laser makes contact with the surface particles they are instantly fused by the 12°C (54°F) rise in temperature.

QUALITY All these layer-building processes work to fine tolerances and are accurate to around ±0.15 mm (0.006 in.). As a result of manufacturing 3D forms in layers, contours are visible on the surface of acute angles.

TYPICAL APPLICATIONS These processes are used to produce aesthetic and functional prototypes and low-volume production runs of parts for all industries.

COST AND SPEED The cost of rapid prototyping is dependent on the volume of the part and build time. Labour costs are moderate, although they depend on the finishing required.

MATERIALS A range of materials are suitable for rapid prototyping, including nylon-based powders, ceramic powders, epoxy resin and certain metal alloys.

ENVIRONMENTAL IMPACTS Most scrap material created during rapid prototyping can be recycled. These processes are an efficient use of energy and material.

Carbon composite impeller (above) The SLS technique is often selected to produce functional prototypes and test models because the materials have similar physical characteristics to injection molded parts. In this case, a composite of carbon fibre strands and nylon powder (50:50) has been selective laser sintered. This material is used in the manufacture of high strength and lightweight parts. Aluminium and nylon powder mixes (50:50) are used for similar high-performance applications.

Nylon engine model (left) Complex and intricate parts can be formed using selective laser sintering. The non-sintered powder forms a 'cake', which encapsulates and supports the model as the build progresses. By contrast, parts made by other layer-building processes need to be supported and undercuts must be tied into the build platform.

1

Following the sintering process (image **1**), the delivery chamber moves up to deliver powder to the roller. This spreads it across the surface of the build area, providing an even layer of powder (image **2**) ready for the next sintering process.

The non-sintered 'cake' encapsulates the parts and has to be carefully brushed away so that individual parts can be removed for cleaning (image **3**). The final part is an exact replica of the computer model, accurate to 150 microns (0.0059 in.) (image **4**).

2

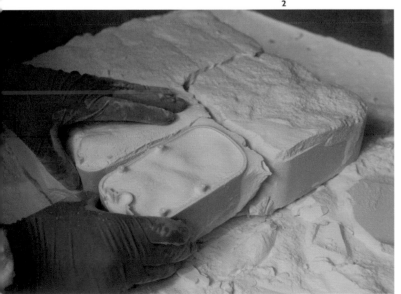

3

4

What is Direct Metal Laser Sintering (DMLS)?

During the sintering process, the delivery chamber rises to dispense powder in the path of the paddle, which spreads a precise layer over the build area. The build platform is incrementally lowered as each layer of metal alloy is sintered onto the surface of the part. The whole process takes place in an inert nitrogen atmosphere at less than 1% oxygen to prevent oxydization of the metal powder during the build.

A considerable amount of heat is generated during this process because a 250 watt CO_2 laser is used to sinter the metal alloy powders.

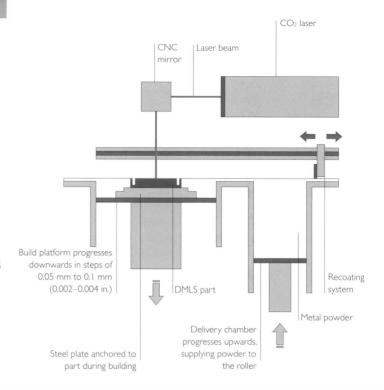

CO₂ laser

CNC mirror

Laser beam

Build platform progresses downwards in steps of 0.05 mm to 0.1 mm (0.002–0.004 in.)

DMLS part

Recoating system

Metal powder

Steel plate anchored to part during building

Delivery chamber progresses upwards, supplying powder to the roller

Functional prototypes DMLS is used to produce functional metal prototypes and low-volume production runs of parts for the automotive, Formula 1, jewelry, medical and nuclear industries.

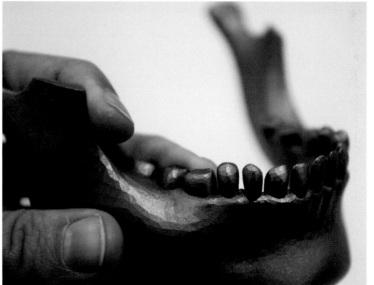

Stainless steel part Stainless steel is used in a variety of medical, aerospace and other engineering applications requiring high hardness, strength and corrosion resistance.

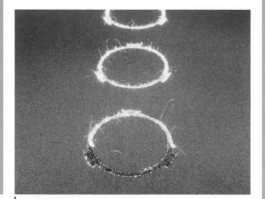

The CO_2 laser is guided across the surface layer of nickel–bronze alloy spheres, 20 microns (0.00078 in.) in diameter (image **1**). After each pass of the laser, a new layer of metal powder is spread over the build area.

Once building is complete, the build platform is raised and the excess powder is brushed away (image **2**) to reveal the parts still attached to the build plate (image **3**). The assembly is removed by hand (image **4**). The parts are eventually removed from the steel build plate by wire EDM (page 56).

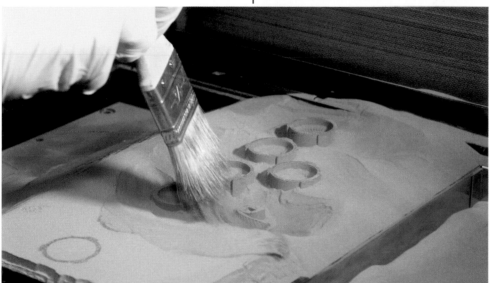

2

3

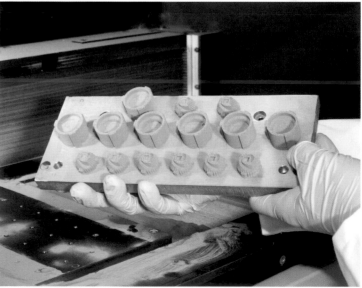

4

What is Stereolithography (SLA)?

The model is built one layer at a time by a UV laser beam directed by a computer-guided mirror onto the surface of the UV sensitive liquid epoxy resin. The UV light precisely solidifies the resin it touches. Each layer is applied by submersion of the build platform into the resin. The paddle sweeps across the surface of the resin with each step downwards, to break the surface tension of the liquid and control layer thickness. The part gradually develops below the surface of the liquid and is kept off the build platform by a support structure. This is made in the same incremental way, prior to building the first layer of the part.

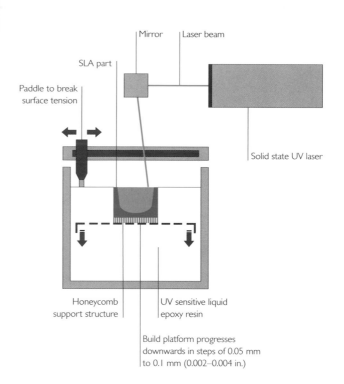

Mirror

Laser beam

SLA part

Paddle to break surface tension

Solid state UV laser

Honeycomb support structure

UV sensitive liquid epoxy resin

Build platform progresses downwards in steps of 0.05 mm to 0.1 mm (0.002–0.004 in.)

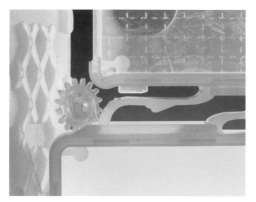

Functional prototypes The high tolerances of the SLA process mean that it is ideal for producing fit-form prototypes that are used to test products before committing to high-volume production. In this case, a part has been built in polyproylene (PP) mimic with live hinges and snap fits.

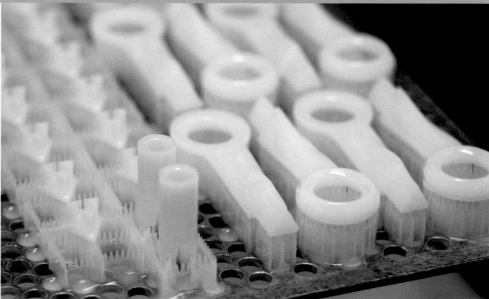

Micro modelling Micro modelling can be used to produce intricate and precise parts (up to 77 × 61 × 230 mm/3.03 × 2.4 × 9.05 in.). This process builds in 25 micron (0.00098 in.) layers, which are almost invisible to the naked eye and so eliminate surface finishing operations. These parts are still attached to the build platform by a support structure, which will be removed when they are cleaned up.

1

2

Case Study

Building a SLA Part

Featured company CRDM www.crdm.co.uk

The SLA parts appear as ghost-like forms in the clear epoxy resin as each pass of the laser fuses another 0.05 mm to 0.1 mm (0.002–0.004 in.) to the preceding layer (image **1**).

The finished parts are removed from the build tank and separated from the build platform (image **2**). An alcohol-based chemical (isopropyl alcohol) is used to clean off the uncured resin liquid and any other contamination and the parts are then fully cured under intensive UV light for one minute (image **3**).

The build strata are just visible in the finished part (image **4**) and can be removed with abrasive blasting, polishing or painting (page 180).

3

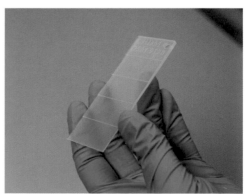

4

Laser Cutting

This is a high-precision CNC process that can be used to cut, etch, engrave and mark a variety of materials including plastic, metal, timber, veneer, paper and card, synthetic marble, flexible magnet, textile and fleece, rubber and certain types of glass and ceramic.

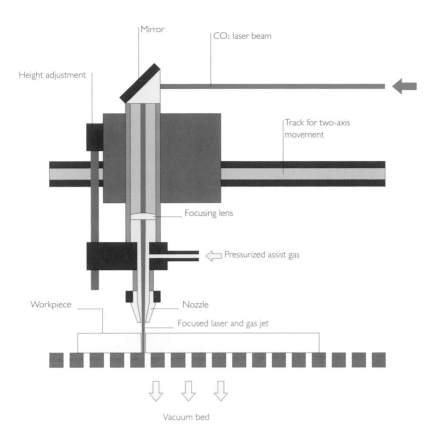

Mirror

CO₂ laser beam

Height adjustment

Track for two-axis movement

Focusing lens

Pressurized assist gas

Workpiece

Nozzle

Focused laser and gas jet

Vacuum bed

Essential Information

VISUAL QUALITY	●●●●●●○○
SPEED	●●●●●○○○
MOLD AND JIG COST	○○●●○○○○
UNIT COST	●●●●●○○○
ENVIRONMENT	●●●●○○○○

Related processes include:
- Laser Engraving
- Laser Scoring

Alternative and competing processes include:
- CNC Engraving
- CNC Machining
- CNC Turret Punching
- Electric Discharge Machining
- Photochemical Machining
- Punching and Blanking
- Water Jet Cutting

What is Laser Cutting?

CO_2 and Nd:YAG laser beams are guided to the cutting nozzle by a series of fixed mirrors. Due to their shorter wavelength, Nd:YAG laser beams can also be guided to the cutting nozzle with flexible fibre optic cores. This means that they can cut along five axes because the head is free to rotate in any direction.

The laser beam is focused through a lens to a fine spot, between 0.1 mm and 1 mm (0.004–0.04 in.). The high concentration beam melts or vaporizes the material on contact.

Notes for Designers

QUALITY Certain materials, like thermoplastics, have a very high-quality surface finish when cut in this way. Laser processes produce perpendicular, smooth, clean cuts in most materials.

TYPICAL APPLICATIONS Applications include furniture, consumer electronics, fashion, signs and trophies, and point of sale.

COST AND SPEED Data is transmitted directly from a CAD file to the laser-cutting machine. Cycle time is rapid but dependent on material thickness. Thicker materials take considerably longer to cut.

MATERIALS This process is ideally suited to cutting thin sheet materials down to 0.2 mm (0.0079 in.); it is possible to cut sheets up to 40 mm (1.57 in.), but thicker materials greatly reduce processing speed. Compatible materials include plastic, metal, timber, veneer, paper and card, synthetic marble, flexible magnet, textile and fleece, rubber and certain types of glass and ceramic.

ENVIRONMENTAL IMPACTS Careful planning will ensure minimal waste, but it is impossible to avoid offcuts that are not suitable for reuse.

Laser-cut plywood architectural model The strength and depth of the CO_2 laser can be controlled to produce a variety of finishes. In this example 1 mm (0.04 in.) thick birch plywood is being cut and scored to form part of an architectural model. The first pass scores surface details laid out in the top layer of the CAD file. Secondly, the laser cuts internal shapes and finally the outside profile. The parts are removed and assembled to form a building façade in relief.

1

Laser Cutting, Raster Engraving and Scoring

Featured company Zone Creations
www.zone-creations.co.uk

The pattern was designed by Ansel Thompson for Vexed Generation in 2005. This series of samples demonstrates the versatility of the laser-cutting process. The laser is used to cut translucent 3 mm (0.118 in.) thick poly methyl methacrylate (PMMA) (image **1**).

The laser leaves a polished edge on PMMA materials and so finishing operations are not required (image **2**).

Raster engraving uses only a small percentage of the laser's power and produces engravings up to 40 microns (0.0016 in.) deep (image **3**).

Scoring produces 'edge glow' effect in the cut detail (image **4**). This is caused by light picked up on the surface of the material being transmitted out through the edges. The scoring acts like an edge and so lights up in the same way.

2

3

4

Water Jet Cutting

A supersonic jet of water, which is typically mixed with abrasives, is used to cut through almost any sheet material from soft foam to titanium. It is a versatile process: intricate and complex profiles can be cut out with a water jet, as well as stainless steel up to 60 mm (2.36 in.) thick.

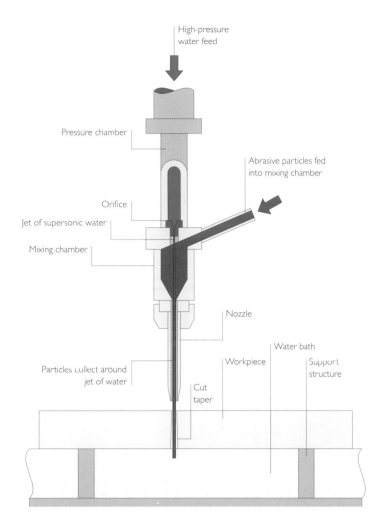

High-pressure
water feed

Pressure chamber

Abrasive particles fed
into mixing chamber

Orifice

Jet of supersonic water

Mixing chamber

Nozzle

Particles collect around
jet of water

Water bath

Workpiece

Support
structure

Cut
taper

What is Water Jet Cutting?

Carried out as either water only cutting or abrasive water jet cutting, water is supplied to the cutting nozzle at very high pressure up to 4,000 bar (60,000 psi). It is forced through a small opening in the 'orifice' (0.1 to 0.25 mm/0.004 to 0.01 in. in diameter).

In abrasive water jet cutting the sharp mineral particles (often garnet) are fed into the mixing chamber and combined with the supersonic water. The abrasive particles create a beam 1 mm (0.04 in.) in diameter, which produces the cutting action.

The high-velocity jet is dissipated by the bath of water below the workpiece. This water is continuously sieved, cleaned and recycled.

QUALITY One of the main advantages of water jet technology is that it is a cold process; therefore it does not produce a heat-affected zone (HAZ), which is most critical in metals. This also means that there is no discolouration along the cut edge and pre-printed or coated materials can be cut this way.

TYPICAL APPLICATIONS As with most new technology, the aerospace and advanced automotive industries were the early adopters. However, this process is now an essential part of many factories.

COST AND SPEED Molds and jigs are not required. Cycle time can be quite slow but depends on the thickness of material and quality of cut. Labour costs are moderate.

MATERIALS Most sheet materials including metal, ceramic, glass, wood, textiles and composites.

ENVIRONMENTAL IMPACTS There are no hazardous materials created in the process or dangerous vapours off-gassed. The water is usually tapped from the mains and is cleaned and recycled for continuous use.

Cut-edge finish Water jet cutting produces a matt surface finish. Pure water jet produces a much cleaner cut than abrasive systems. The sharp particles used in abrasive water jet cutting vary in size much like sandpaper (120, 80 and 50). Different grit sizes affect the quality of the surface finish; finer grit (higher number) is slower and produces a higher quality surface finish.

Intricate shapes Water jet cutting does not create stresses in the workpiece, so small, intricate and complex profiles are possible. Most glass materials can be cut, between 0.5 mm and 10 cm thick (0.02–3.94 in.). The hardness of the material will determine the maximum thickness.

Water Jet Cutting Glass

Featured company Instrument Glasses
www.instrumentglasses.com

This process is CNC, so the cutting data is generated from a CAD file. The cutting nozzle (image **1**) progresses slowly over the 25 mm (0.98 in.) plate glass. As it progresses, the operator inserts wedges to support the part being cut out (image **2**).

After cutting, the part is carefully removed (image **3**) and prepared for finishing (image **4**). The cut edge is improved by polishing (page 170), which takes place on a diamond encrusted polishing wheel, or flaming.

1

3

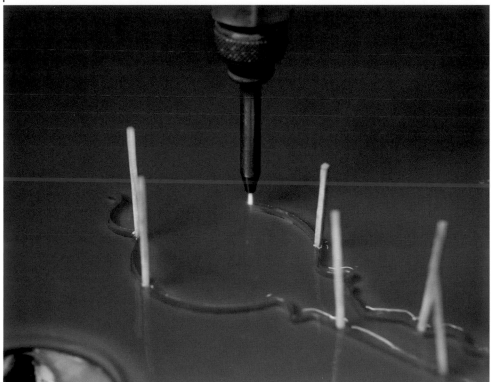

2

4

Photochemical Machining

This chemical process is used to mill and machine thin sheet metals. It is also known as chemical blanking and photofabrication. Decorative chemical cutting is known as photo etching. Unprotected metal, which has not been masked, is dissolved to produce a cut-out profile or engraving.

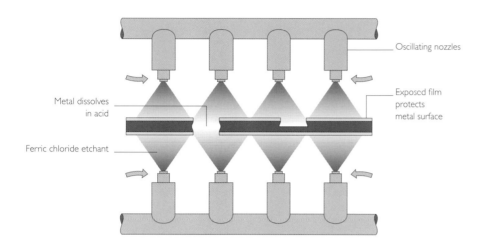

Oscillating nozzles

Metal dissolves
in acid

Exposed film
protects
metal surface

Ferric chloride etchant

Related processes include:
• Engraving
• Profiling

Alternative and competing processes include:
• CNC Engraving
• CNC Machining
• CNC Turret Punching
• Laser Cutting
• Photo Etching
• Punching and Blanking

What is Photochemical Machining?

A coating of resist film is applied to both sides of the workpiece and exposed, under a negative, to ultraviolet light. Afterwards, the unexposed areas of the photosensitive resist film are chemically developed away. This process exposes the areas of the metal that are to be etched.

The metal sheet is passed under a series of oscillating nozzles that apply the chemical etch. The oscillation ensures that plenty of oxygen is mixed with the acid to accelerate the process. Finally, the protective polymer film is removed from the metalwork to reveal the finished etching.

QUALITY This process produces an edge finish free from burrs, and is accurate to within 10% of the material thickness. The surface finish is matt, but can be polished.

TYPICAL APPLICATIONS The technical aspects are utilized in the aerospace, automotive and electronics industries. Other products include model-making nets, control panels, meshes and jewelry.

COST AND SPEED The only tooling required is a negative that can be printed directly from CAD data or a graphics software file. Cycle time and labour costs are moderate.

MATERIALS Metals including stainless steel, mild steel, aluminium, copper, brass, nickel, tin and silver. Of these, aluminium is the easiest, and stainless steel is the hardest and so takes longer to etch.

Glass, mirror, porcelain and ceramic are also suitable for photo etching, although different types of photo-resist and etching chemical are required.

ENVIRONMENTAL IMPACTS The chemical used to etch the metal is one third ferric chloride. Caustic soda is needed to remove spent protective film. Both of these chemicals are harmful and operators must wear protective clothing.

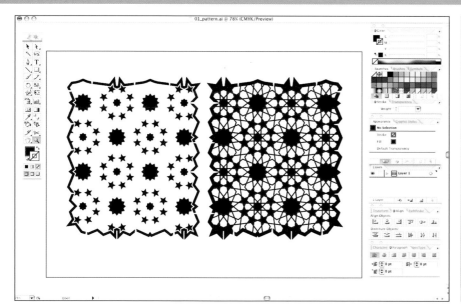
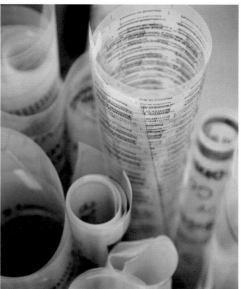

Negatives The design for the negatives is prepared using graphics software, which shows how the chemical machining process works. The left-hand drawing is the reverse and the right-hand drawing is the front of the workpiece. The two sides are etched simultaneously, so the areas that are blacked out on both sides will be cut through (profiled). Areas that are black only on one side will be half etched. The negatives are printed onto acetate.

Case Study

Photochemical Machining a Brass Screen

Featured company Aspect Signs & Engraving
www.aspectsigns.com

The workpiece is coated with photosensitive film and placed between the acetate negatives (image **1**). The assembly is placed under a vacuum and exposed to UV light on both sides (image **2**). Unexposed photosensitive film is washed off in a developing process.

After the first stage of chemical machining (image **3**), the process is repeated until the chemical has etched through the entire thickness of the material. The geometric pattern is simultaneously etched and profiled (image **4**).

1

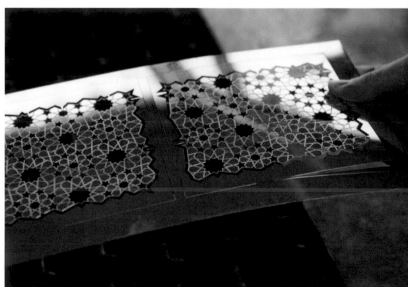

2

4

3

Press Braking

This simple and versatile technique is utilized to bend sheet-metal profiles for prototypes and low-volume production. Combined with cutting and joining techniques, a range of geometries can be formed including simple bends, continuous profiles and enclosures. It is also referred to as brake forming.

Hydraulic ram

Punch

Workpiece (blank)

Die

Stage 1: Load

Stage 2: Air bending

Bottom bending

Gooseneck bending

Essential Information

VISUAL QUALITY	●●○○○○○○
SPEED	●●●●●○○○
MOLD AND JIG COST	●●○○○○○○
UNIT COST	●●●●○○○○
ENVIRONMENT	●●○○○○○○

Alternative and competing processes include:
• Metal Extrusion
• Metal Press Forming
• Roll Forming

What is Press Braking?

In stage one, the workpiece is loaded onto the die. In stage two, vertical pressure is applied by the hydraulic ram so the punch forces the part to bend. Each bend takes only a few seconds.

The geometry of the bend determines the type of punch and die that are used; there are many different types, including air-bending dies, V-dies for bottom ending, gooseneck dies, acute-angle dies and rotary dies.

Air bending is used for most general work, while bottom bending with matched dies (also called V-die bending or coining) is reserved for high-precision metalwork. Gooseneck dies are for bending re-entrant angles that cannot be accessed by a conventional punch.

Notes for Designers

QUALITY Applying a bend to a sheet of material increases its strength. Machines are computer guided which means they are precise to within at least 0.01 mm (0.0004 in.).

TYPICAL APPLICATIONS Applications include lorry sidings, architectural metalwork, interiors, kitchens, furniture and lighting, prototypes and general structural metalwork.

COST AND SPEED Standard tooling is suitable for most applications. Specialized tooling will increase the unit price, depending on the size and complexity of the bend. Cycle time is up to six bends per minute on modern computer-guided equipment. Labour costs are high for manual operations.

MATERIALS Almost all metals can be formed using press braking, including steel, aluminium, copper and titanium.

ENVIRONMENTAL IMPACTS Bending is an efficient use of materials and energy. There is no scrap in the bending operation, although there may be scrap produced in the preparation and in subsequent finishing operations.

Press braking punches Many different shapes of bend can be formed, without investment in tooling, using standard punches like these.

Case Study

Press Braking an Aluminium Enclosure

Featured company Cove Industries
www.cove-industries.co.uk

Aluminium blanks are prepared for press braking by CNC turret punching, guillotining or laser cutting. In this case study, the aluminium blanks were cut out using a turret punch (page 48) (image **1**).

Gooseneck bending is used when it would not be possible to form the second bend with a conventional punch. The metal blank is inserted against a computer-guided stop (image **2**), which ensures that the part is located precisely prior to bending. The downstroke of the punch is smooth to avoid stressing the material unnecessarily and takes only a couple of seconds (image **3**).

The formed joints are ready for welding (page 134) and polishing (page 170) (image **4**).

Centrifugal Casting

Centrifugal casting covers a range of spinning processes used to shape materials in their liquid state. By spinning at high speed, the material is forced to flow through the mold cavities. Because mold costs are very low, this process is used to prototype as well as mass produce millions of parts.

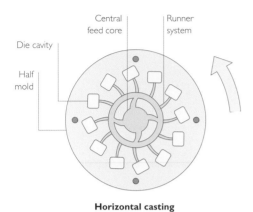

Die cavity
Central feed core
Half mold
Runner system

Horizontal casting with multi-cavity tool

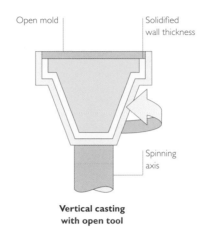

Open mold
Solidified wall thickness
Spinning axis

Vertical casting with open tool

Essential Information

VISUAL QUALITY	●●●●○○○○
SPEED	●●●●○○○○
MOLD AND JIG COST	●●○○○○○○
UNIT COST	●●●●○○○○
ENVIRONMENT	●●●○○○○○

Related processes include:
• Horizontal Casting
• Vertical Casting

Alternative and competing processes include:
• Die Casting
• Forging
• Metal Injection Molding
• Rapid Prototyping
• Sand Casting

What is Centrifugal Casting?

The horizontal casting technique shown above uses molds made from silicone or metal. Molten material is fed along the central feed core, which is in the upright position. As the mold spins the metal is forced along the runner system and into the die cavities. Flash forms between the meeting point of the two mold halves. Runners can be integrated into the mold to encourage air flow out of the die cavity.

Vertical casting methods are similar to rotation molding. The difference is that in centrifugal casting the mold rotates around a single axis, whereas rotation molding tools are spun around two axes or more. In operation, a skin of molten material forms over the inside surface of the mold to form sheet or hollow geometries.

QUALITY Surface details, complex shapes and thin wall sections are reproduced very well. Metals produced in silicone-mold centrifugal casting have a low melting point. Therefore, they will not be as strong and resilient as metals formed in other ways.

TYPICAL APPLICATIONS Prototypes and production runs for jewelry, bathroom fittings and architectural models.

COST AND SPEED Silicone mold costs are very low. Metal molds are more expensive, but still relatively inexpensive for their size. The cycle time for low melting point metals and plastics ranges from 30 seconds to five minutes. Using multiple cavity molds reduces unit costs.

MATERIALS Silicone molds can be used to cast some plastics, including polyurethane. Metals include white metal, pewter and zinc.

ENVIRONMENTAL IMPACTS Most scrap can be directly recycled. White metal and pewter are alloys of lead. The exception to this rule is British Standard pewter, which contains no lead.

Scale alloy wheel Very small and intricate parts are feasible with centrifugal casting, such as this 12:1 scale model of an alloy wheel in pewter, which is part of a model Raleigh car.

Multi-cavity mold Silicone-mold centrifugal casting can produce from one to 100 parts in a single cycle. This multi-cavity silicone split mold example produces 35 metal parts simultaneously.

Case Study

Centrifugal Casting a Scale Model

Featured company CMA Moldform Limited
www.cmamoldform.co.uk

This cast pewter part is quite large (image **1**) and so each mold can produce only one part at a time. The silicone mold is assembled with cores (image **2**) and the two halves of the mold are brought together and located with pimples and matching recesses on the interface.

The mold is clamped onto the spinning table and sealed. The mold is spun at high speed and molten pewter is poured into the central feed core (image **3**) and enters the mold where it is pushed through the runner system by centrifugal force.

The mold halves are separated and the cast metal part is removed (image **4**).

1

2

3

4

Joining Technology

2

Arc Welding

Arc welding is an essential part of the fabrication process, used extensively in the metalworking industries. Arc-welding processes can only be used to join metals because they rely on the formation of an electric arc between the workpiece and electrode to produce heat. The most common types are MMA, MIG and TIG.

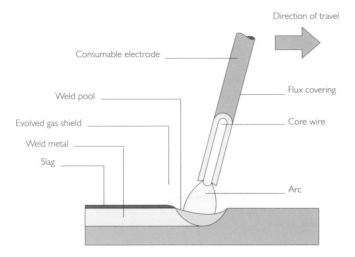

Direction of travel

Consumable electrode

Flux covering

Weld pool

Core wire

Evolved gas shield

Weld metal

Slag

Arc

Essential Information

VISUAL QUALITY	●●●●●○○○
SPEED	●●○○○○○○
MOLD AND JIG COST	○○○○○○○○
UNIT COST	●●●●○○○○
ENVIRONMENT	●●●○○○○○

Related processes include:
• Manual Metal Arc (MMA)
• Metal Inert Gas (MIG)
• Tungsten Inert Gas (TIG)

Alternative and competing processes include:
• Friction Welding
• Power Beam Welding
• Resistance Welding
• Soldering and Brazing
• Ultrasonic Welding

What is Manual Metal Arc (MMA) Welding?

MMA welding, also known as stick welding, has been in use since the late 1800s, but has seen major development in the last 60 years. Modern welding techniques use a coated electrode. The coating (flux) melts during welding to form the protective gas shield and slag.

MMA welding can only produce short lengths of weld before the electrode needs to be replaced. This is time consuming and stresses the welded joint because the temperature is uneven. The slag that builds up on top of the weld bead has to be removed before another welding pass can take place.

Notes for Designers

QUALITY The quality of manual arc welding is largely dependent on operator skill. It is possible to form precise and clean weld beads with all manual techniques.

TYPICAL APPLICATIONS These welding processes are used extensively in the metalworking industries, including the furniture, automotive and construction industries.

COST AND SPEED Jigs may be required to locate the parts accurately during welding. Cycle time is slow for MMA welding, MIG welding is rapid, especially if automated, and TIG welding falls somewhere inbetween. Labour costs are high for manual operation due to the level of skill required.

MATERIALS MMA welding is generally limited to steel, iron, nickel and copper. TIG is widely used on carbon steel, stainless steel, titanium and aluminium. MIG welding is commonly used to join steels, aluminium and magnesium.

ENVIRONMENTAL IMPACTS A constant flow of electricity generates a great deal of heat and there is very little heat insulation, so it is relatively inefficient. There is limited scrap.

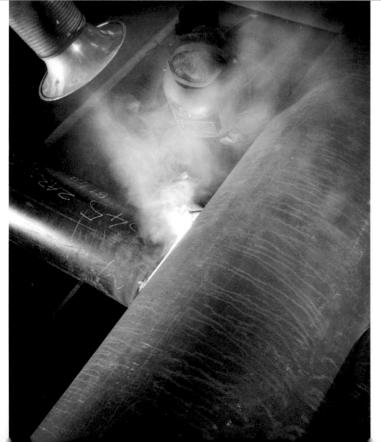

On the spot welding MMA welding is the most portable of the processes, requiring relatively little equipment; it is used a great deal in the construction industry and for other site applications. The equipment can be used in horizontal, vertical and inverted positions, making this versatile for manual operations.

1

Case Study

Manual Metal Arc (MMA) Welding Steelwork

Featured company TWI www.twi.co.uk

The operator is welding a steel plate assembly (image **1**). The weld is formed by the electrode and workpiece heating and melting to form a weld pool, which rapidly solidifies to form a bead of weld metal (image **2**). A cloud of shielding gas and a layer of slag, which must be removed after each welding pass, protects the molten weld pool from the atmosphere and encourages the formation of a 'sound' joint.

The electrode is consumed as the operator progresses along the joint and so has to be replaced (image **3**).

The appearance and strength of the weld is largely dependent on the skill of the operator (image **4**).

2

3

4

What is Metal Inert Gas (MIG) Welding?

MIG welding is principally the same as MMA welding; the weld is made by forming an arc between the electrode and the workpiece and is protected by a plume of inert gas. MIG welding distinguishes itself from MMA welding through higher productivity rates, greater flexibility and suitability for automation.

The gas shield performs a number of functions, including aiding the formation of the arc plasma, stabilizing the arc on the workpiece and encouraging the transfer of molten electrode to the weld pool. Generally, the gas is a mixture of argon, oxygen and carbon dioxide.

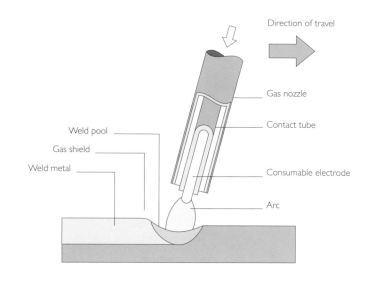

Direction of travel

Gas nozzle

Contact tube

Weld pool

Gas shield

Weld metal

Consumable electrode

Arc

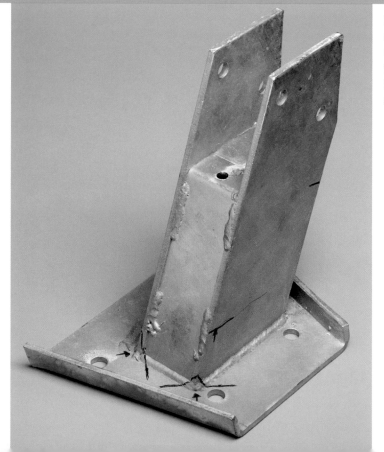

MIG welded aluminium MIG welding accounts for about half of all welding operations and is used in many industries. It is used extensively in the automotive industry because it is rapid and produces clean welds in steel, aluminium and magnesium. This is an example of a MIG welded aluminium assembly.

1

Metal Inert Gas (MIG) Welding Steelwork

Featured company TWI www.twi.co.uk

Here MIG welding is being utilized to seal the end of a steel pipe (image **1**). Similar to MMA welding, MIG welding equipment is highly portable. The difference is that MIG welding uses an electrode continuously fed from a spool (image **2**) and the shielding gas is supplied separately.

Localized heating between the electrode and workpiece causes the surfaces to coalesce and form a strong joint (image **3**), which is supported by the filler material. Care must be taken in the design process to ensure a strong joint.

2

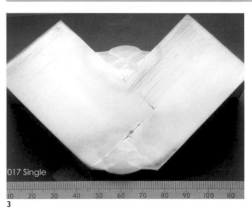

017 Single

3

What is Tungsten Inert Gas (TIG) Welding?

TIG welding is a precise and high-quality welding process. It is ideal for thin sheet materials and precise and intricate work. The main distinction is that TIG welding does not use a consumable electrode; instead, it has a pointed tungsten electrode. The weld pool is protected by a shielding gas and filler material can be used to increase deposition rates and for thicker materials.

   The shielding gas is usually helium, argon or a mixture of both. Argon is the most common and is used for TIG welding steels, aluminium and titanium.

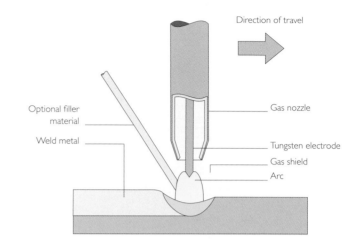

Direction of travel

Optional filler material

Weld metal

Gas nozzle

Tungsten electrode

Gas shield

Arc

TIG welding titanium TIG is widely used on carbon steel, stainless steel and aluminium, and it is the main process for joining titanium.

1

Case Study

Tungsten Inert Gas (TIG) Welding

Featured company TWI www.twi.co.uk

The quality and slower speeds of TIG welding make it ideally suited to precise and demanding applications. It may be manual or fully automated. An arc is formed between the tungsten electrode and workpiece, and the filler material melts into the weld pool (images **1** and **2**).

As with all thermal metal fabrication processes, TIG welding produces a heat-affected zone (HAZ) (image **3**). The problems are caused by a change in structure which leads to a change in the properties of the material and a susceptibility to cracking. The visual affects of the HAZ can generally be removed during finishing.

2

NWA2 - (ii)

3

Soldering and Brazing

These processes have been used in metalworking for centuries. Both are used to form permanent joints by melting a filler material into the joint. The difference between the processes is the melting point of the filler materials, which is lower for soldering than it is for brazing.

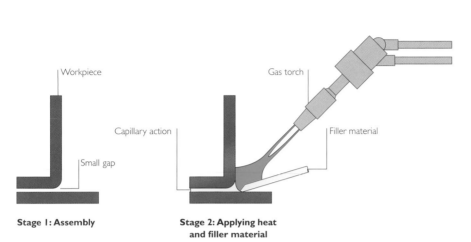

Workpiece

Capillary action

Small gap

Stage 1: Assembly

Gas torch

Filler material

Stage 2: Applying heat and filler material

Essential Information

VISUAL QUALITY	●●●●○○○○
SPEED	●●○○○○○○
MOLD AND JIG COST	●○○○○○○○
UNIT COST	●●●●●○○○
ENVIRONMENT	●●●○○○○○

Alternative and competing processes include:
- Arc Welding
- Resistance Welding

What is Soldering and Brazing?

There are many different techniques, but the basic principle of soldering and brazing is that the workpiece is heated to above the melting point of the filler material. At this point the filler becomes molten and is drawn into the joint by capillary action. The liquid metal filler forms a metallurgical bond with the workpiece to create a joint that is as strong as the filler material itself.

The filler material is typically an alloy of silver, brass, tin, copper or nickel, or a combination. The choice of filler material is determined by the workpiece material because it has to be metallurgically compatible.

Notes for Designers

QUALITY Brazing is generally stronger than soldering because the filler material has a higher melting point than solder. The finish of the joint bead is usually satisfactory without the need for significant grinding. And even though brazing is usually carried out with brass filler, the colour can be adjusted to suit the workpiece material.

TYPICAL APPLICATIONS Applications include jewelry, plumbing, silverware, bicycle frames and watches.

COST AND SPEED Specially designed jigs may be required to secure the part during soldering and brazing. Cycle time is rapid and ranges from one to ten minutes for most torch applications. Labour costs are generally low.

MATERIALS Most metals and ceramics can be joined using these techniques. Metals include aluminium, copper, carbon steel, stainless steel, nickel, titanium and metal matrix composites.

ENVIRONMENTAL IMPACTS Soldering and brazing operate at lower temperatures than arc welding. There are very few rejects because faulty parts can be dismantled and reassembled.

Case Study

Brazing the Bombé Milk Jug

Featured company Alessi www.alessi.com

I

The Bombé milk jug was designed by Carlo Alessi in 1945 and is still in production today (image **1**). Even though it would be quicker and more cost effective to resistance weld this joint, brazing is still used to maintain the integrity of the original design.

The joint is coated with flux paste (image **2**). The two parts are mounted into a jig (image **3**). The brazing process is very rapid, lasting 30 seconds or so (image **4**). Afterwards, the finished jug is lightly polished and cleaned (image **5**).

Resistance Welding

These are rapid techniques used to join sheet metal. High voltage, concentrated between two electrodes, causes the metal to heat and coalesce. Spot and projection welding are used for assembly operations and seam welding is used to produce a series of overlapping weld nuggets which form a hermetic seal.

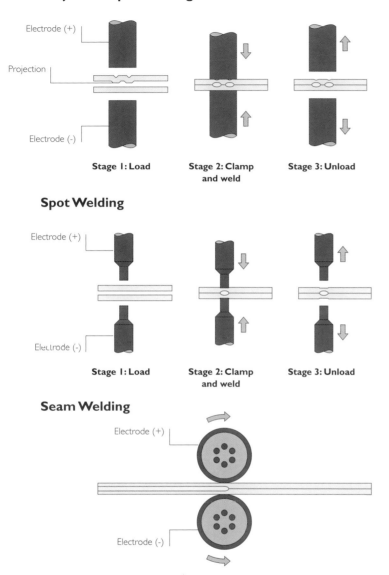

Projection Spot Welding

Electrode (+)

Projection

Electrode (-)

Stage 1: Load   **Stage 2: Clamp and weld**   **Stage 3: Unload**

Spot Welding

Electrode (+)

Electrode (-)

Stage 1: Load   **Stage 2: Clamp and weld**   **Stage 3: Unload**

Seam Welding

Electrode (+)

Electrode (-)

What is Resistance Welding?

In projection spot welding, the weld zone is localized. This can be done in two ways: either projections are embossed onto one side of the joint, or a metal insert is used. This process is capable of producing multiple welds simultaneously because unlike spot welding the voltage is directed by the projection or insert. The electrodes do not determine the size and shape of the weld. Therefore, they can have a large surface area that will not wear as rapidly as spot-welding electrodes.

Notes for Designers

QUALITY Weld quality is consistently high. Joints have high shear strength, but peel strength can be limited with small and localized weld nuggets.

TYPICAL APPLICATIONS Applications are widespread, including the automotive, construction, furniture, appliance and consumer electronic industries.

COST AND SPEED Specially designed jigs may be required to weld contoured surfaces, but is generally small and not expensive. Cycle time is rapid.

MATERIALS Most metals can be joined by resistance welding, including carbon steels, stainless steels, nickels, aluminium, titanium and copper alloys.

ENVIRONMENTAL IMPACTS No consumables (such as flux, filler or shielding gas) are required for most resistance-welding processes. Water is sometimes used to cool the copper electrodes, but this is usually recycled continuously without waste.

Spot welding stainless steel mesh These are simple and versatile processes. They are relatively inexpensive and can be used on a range of materials. In this case, spot welding has being used to join stainless steel mesh with a hand-held welder.

Heavy duty hand-held welding guns like this Portable Welder are used for general sheet-metal assembly work. Sophisticated computer-guided robotic systems are used for high-volume welding operations.

Case Study

Projection Spot Welding

The metal ring is placed onto a lower electrode, which locates it to ensure repeatable joints (image **1**). The second part has protrusions that localize the voltage (image **2**). It is clamped into the upper electrode. Projection welding takes a second or so (image **3**). Both welds are formed simultaneously and are full strength almost immediately. The finished parts are stacked (image **4**).

Wood Joinery

Joinery is an essential part of furniture making. Craft and industry have combined over the years and as a result a standard selection of joint configurations have emerged. These include butt, lap, mitre, housing, mortise and tenon, M-joint, scarf, tongue and groove, comb, finger and dovetail.

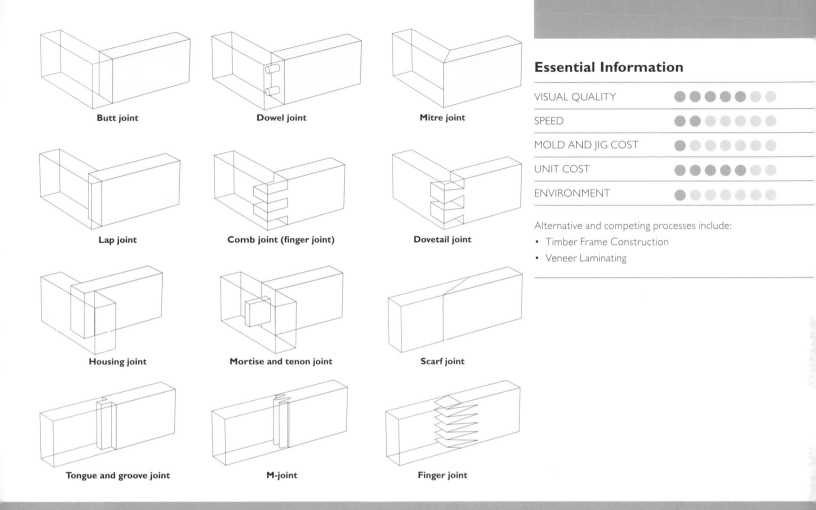

Butt joint

Dowel joint

Mitre joint

Lap joint

Comb joint (finger joint)

Dovetail joint

Housing joint

Mortise and tenon joint

Scarf joint

Tongue and groove joint

M-joint

Finger joint

Essential Information

VISUAL QUALITY	●●●●●●○○
SPEED	●●○○○○○○
MOLD AND JIG COST	●○○○○○○○
UNIT COST	●●●●●○○○
ENVIRONMENT	●○○○○○○○

Alternative and competing processes include:
* Timber Frame Construction
* Veneer Laminating

What is Wood Joinery?

The diagram illustrates the most common joint types. They include handmade and machine-made configurations, which are used in furniture construction, house building and interior structures.

There are four main types of adhesive: urea polyvinyl acetate (PVA), urea formaldehyde (UF), two-part epoxies and polyurethane (PUR). PVA and UF resins are the least expensive and most widely used. PVA is water based and non-toxic, and excess can be cleaned with a wet cloth. PUR and two-part epoxies can be used to join wood to other materials, such as metal, plastic or ceramic, and are waterproof and suitable for exterior use. They are rigid and so restrict the movement of the joint more than PVA.

Notes for Designers

QUALITY Products made from wood have unique characteristics associated with visual patterns (growth rings), smell, touch, sound and warmth. The quality of the joint is very much dependent on skill.

TYPICAL APPLICATIONS Joinery is used in woodworking industries, including furniture and cabinetmaking, construction, interiors, boatbuilding and patternmaking.

COST AND SPEED Jigs are not usually required. Cycle time is totally dependent on the complexity of the job. Labour costs tend to be quite high due to the level of skill required.

MATERIALS The most suitable wood for joinery is solid timbers, including oak, ash, beech, pine, maple, walnut and birch.

ENVIRONMENTAL IMPACTS Wood has many environmental benefits, especially if it is sourced from renewable forests. Timber is biodegradable, can be reused or recycled and does not cause any pollution.

Joints tend to be formed by cutting, so waste is unavoidable. Dust, shavings and wood chips are often burnt to reclaim energy in the form of heating.

Case Study

Mitred and Biscuit-reinforced Butt Joints in a Bedside Table

Featured company Windmill Furniture
www.windmillfurniture.com

This is a simple, veneered bedside table. The four sides of the product are lipped with solid oak edges, veneered, cut to size and mitred. Grooves are cut for the biscuit-reinforced butt joints.

Prior to assembly, the biscuits (image **1**) are placed into pre-cut grooves on the butt joint, and glue is applied to all of the joints (image **2**).

The four sides are assembled with the tape still in place (image **3**). This keeps the joint tight and acts like a clamp (image **4**).

3

1

2

4

1

Case Study

Mortise and Tenon and M-joints in the Home Table

Featured company Isokon Plus www.isokonplus.com

This is the Home Table, which was designed by BarberOsgerby in 2000. It is produced from solid oak and utilizes a range of different joints (image **1**). The legs and table frame are joined with a mortise and tenon, the traditional and strongest joint for this application (images **2** and **3**). The top is made by joining solid planks of oak with M-joints (images **4** and **5**).

2

3

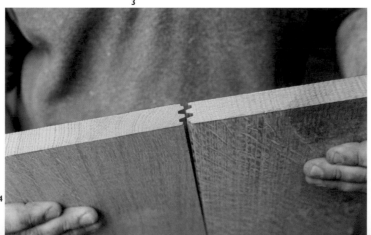

4

5

Case Study

Comb Jointed Tray

Featured company Windmill Furniture
www.windmillfurniture.com

Comb joints are traditionally used to join the sides of trays and drawers. The joint is cut by spindle molder with a set of cutters separated by matched spacers. Both sides are cut with the same set up to ensure a perfect fit (image **1**). It can be assembled by hand (image **2**).

The base of the tray, which is located in a housing joint between the four sides, holds it all square and the finished product is lacquered (image **3**).

Case Study

Dowelled Butt Joints in a Table Drawer

Featured company Windmill Furniture
www.windmillfurniture.com

This case study illustrates dowels used to strengthen butt-joint configurations. The parts are drilled with the holes set apart to exact measurements (image **1**). This is often carried out on a drill with two heads, which are set exactly the same distance apart for both sides of the joint.

The dowels are made from beech or birch because they are suitably hard materials (image **2**). They are inserted into the joint with glue and hammered into place (image **3**).

1

2

Case Study

Housing Joints in the Donkey

Featured company Isokon Plus www.isokonplus.com

This version of the Donkey was designed by Egon Riss in 1939. It is made from birch plywood (image **1**). Housing joints are a simple and strong way to fix the shelves into the end caps (image **2**). Using this joint in this way means that the product can be assembled and glued in a single operation. The end caps are clamped together which applies even pressure to all the joints.

1

2

3

Case Study

Decorative Inlay

Featured company Windmill Furniture
www.windmillfurniture.com

This simple form of wood inlay is used to separate two veneers on a tabletop visually (image **1**). The inner veneer is bird's-eye maple and the outer veneer is plain maple. This type of decorative inlay is made up of layers of exotic hardwoods and fruitwoods, which are cut into strips (image **2**). The groove is cut by a router and the strips of inlay are bonded in with UF adhesive (image **3**).

Finishing Technology

3

Electroplating

This process is used to apply a thin film of metal to another metal surface for functional and visual effects. Electroplated metals benefit from a combination of the properties of the two materials. For example, silver-plated brass combines the strength and reduced cost of brass with the long-lasting lustre of silver.

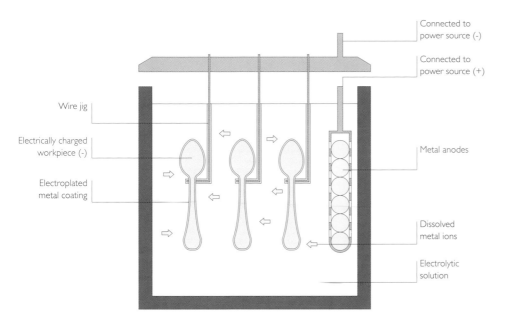

Connected to
power source (-)

Connected to
power source (+)

Wire jig

Electrically charged
workpiece (-)

Electroplated
metal coating

Metal anodes

Dissolved
metal ions

Electrolytic
solution

What is Electroplating?

Electroplating occurs in an electrolytic solution of the plating metal held in suspension in ionic form. When the workpiece is submerged and connected to a DC current, a thin film of electroplating forms on its surface. The rate of deposition depends on the temperature and chemical content of the electrolyte.

As the thickness of electroplating builds up on the surface of the workpiece the ionic content of the electrolyte is replenished by dissolution of the metal anodes. The anodes are suspended in the electrolyte in a perforated container.

Notes for Designers

QUALITY The quality of surface finish is largely dependent on the surface finish of the workpiece prior to electroplating. The whole process is computer controlled to ensure maximum precision and quality of surface finish.

TYPICAL APPLICATIONS This process is used a great deal by jewellers and silversmiths. Examples include rings, watches and bracelets. Tableware includes beakers, goblets, plates and trays.

COST AND SPEED Wire jigs are required to support the parts during electroplating. Cycle time depends on the rate of deposition.

It is approximately 25 microns (0.00098 in.) per hour for silver and up to 250 microns (0.0098 in.) per hour for nickel. Labour costs are moderate to high depending on the application.

MATERIALS Most metals can be electroplated. However, metals combine with different levels of purity and efficiency.

ENVIRONMENTAL IMPACTS Many hazardous chemicals are used in all of the electroplating processes. They are carefully controlled with extraction and filtration to ensure minimal environmental impact.

Gold and silver plating Electroplating can produce the look, feel and benefits of one metal on the surface of another, allowing parts to be formed in materials that are less expensive or have suitable properties for the application. Electroplating then provides them with a metal skin that possesses all the desirable aesthetic qualities.

Gold and silver are inert and suitable for all types of products including beakers, bowls, jewelry and even medical implants.

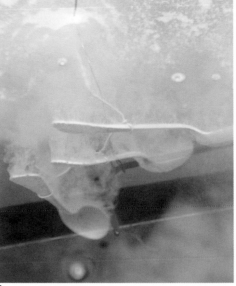

Silver Electroplating Nickel-silver Cutlery

Featured company BJS Company www.bjsco.com

These nickel-silver spoons are polished to a high gloss in preparation for silver electroplating (image **1**). They are immersed in a series of cleaning solutions, including a dilute cyanide solution, in which the surfaces of the spoons can be seen fizzing (image **2**).

In this case 25 microns (0.00098 in.) of silver has been electroplated onto the surface (image **3**).

The electroplated parts are cleaned and dried (image **4**). The surface is improved and finished with a very fine iron-powder polishing compound in a buffing process known as 'colouring over' to produce a highly reflective finish (image **5**).

1

2

3

4

5

Flocking

Short and densely packed fibres are bonded to the surface of the workpiece, standing on end, to create a soft and vividly coloured finish that feels like velvet. The short lengths of fibre are drawn to the surface of the workpiece electrostatically and are permanently bonded in place.

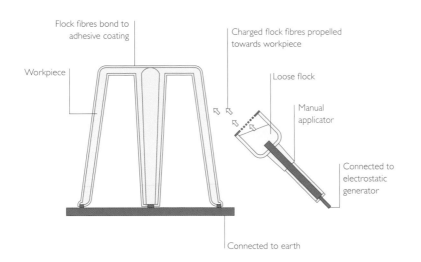

Flock fibres bond to
adhesive coating

Workpiece

Charged flock fibres propelled
towards workpiece

Loose flock

Manual
applicator

Connected to
electrostatic
generator

Connected to earth

Essential Information

VISUAL QUALITY	●●●●●●○○
SPEED	●●●●●○●○
MOLD AND JIG COST	●●●●●●○○
UNIT COST	●●●●●○○○
ENVIRONMENT	●●●●●●○○

Alternative and competing processes include:
• Spray Painting

What is Electrostatic Flocking?

The surface of the workpiece is coated with a conductive adhesive, which is built up to a sufficient thickness to support the fibres once they are applied. The fibres are coated to increase their conductivity.

Flocking is an electrostatic process. The workpiece is grounded and the fibres are charged with high voltage – 40,000–80,000 volts depending on the application – as they leave the applicator. This potential difference draws the fibres towards the surface of the workpiece where they penetrate the layer of adhesive and stand perpendicular to the surface.

As the surface is gradually covered the electrically charged fibres are drawn to the areas that are at earth, which ensures that a dense and even coating is built up.

Notes for Designers

QUALITY Colour is rich, uniform and matt. The softness varies according to the type and length of fibre.

TYPICAL APPLICATIONS Flocking is used for decorative and functional purposes such as automotive applications (glove compartments and interior linings), military parts (anti-reflective), textiles (hats and t-shirts) and products (toys, furniture and lighting).

COST AND SPEED Jigs may be required. Cycle time depends on the size of the part and curing time of the adhesive. Labour costs are moderate for low-volume parts. High-volume applications are typically automated.

MATERIALS Nylon is the most common flocking fibre, however it is possible to flock with other synthetic and natural fibres from 0.2–10 mm (0.0079–0.4 in.) long. For example, rayon tends to be used for textile applications because it is better for skin contact. Almost any material can be flocked.

ENVIRONMENTAL IMPACTS Chemicals are used and dust produced during flocking, so suitable facemasks and breathing equipment are required. Excess flock fibres can be recycled for future use. The surface finish can last for many years, depending on application, and it is possible to retouch damaged coatings.

Flock swatches Flock is available in almost any colour. Mixing different colours and lengths is possible. For example, a composite flock is used to produce a more lifelike fur coating for full-scale models of animals.

Flocking with stencils Stencils applied by vinyl are used to flock graphics and patterns.

Automotive interior lining Flock is used extensively in the automotive industry for functional reasons such as to reduce vibration, noise, condensation and glare (black flock is non-reflective and so is ideal for dashboards, for example).

1

2

3

4

5

Case Study

Flocking a Plastic Stool

Featured company Thomas & Vines Ltd
www.flocking.co.uk

The stool, designed by Rob Thompson in 2003, is flocked with blue trilobal nylon fibres (image **1**). Unlike conventional fibres that have a round profile, trilobal fibres are triangular and so shimmer in the light.

The surface is prepared by spray coating (image **2**) with a suitable adhesive, which in this case is polyurethane resin (PUR).

The nylon fibres are loaded into the applicator (image **3**). The fibres are negatively charged as they leave the applicator, which propels them towards the surface of the electrically grounded workpiece (image **4**). A dense flock coating is achieved within five minutes or so (image **5**).

Metal Patination

A protective oxide layer forms on the surface of copper alloys. It is naturally dark brown or green and can be accelerated and enhanced with artificial patination. This technique uses chemicals and heat to produce a wide range of colours including white, black, red, silver, green and brown.

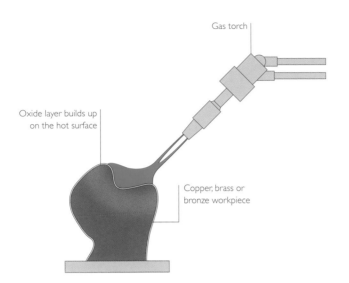

Gas torch

Oxide layer builds up
on the hot surface

Copper, brass or
bronze workpiece

Essential Information

VISUAL QUALITY	●●●●●●●
SPEED	●●●●●●●
MOLD AND JIG COST	●●●●●●●
UNIT COST	●●●●●●●
ENVIRONMENT	●●●●●●●

Alternative and competing processes include:
- Galvanizing
- Polishing
- Spray Painting

What is Artificial Patination?

Artificial patination comprises three main processes: cleaning and preparation, patination and applying a protective wax or oil.

The chemicals are mixed with water and applied evenly to the surface of the metal. Care is taken when stippling with a brush because drips and spills are visible in the final piece.

There are hot and cold techniques. Applying heat accelerates the process and makes the colour and effect visible more quickly. Layers of colour can be built up in stages. Parts that have been patinated cold are typically left over night to allow the colour to develop.

The chemicals react with the surface of the metal to form the thin layer of oxide. This is integral to the base metal and very durable.

QUALITY This technique builds up the naturally occurring oxide layer on the surface of the metal. The film is light, hard, protective and self-healing. The colour and effects are largely dependent on the skill of the patineur and change gradually over time.

TYPICAL APPLICATIONS This technique is typically used for outdoor architectural and artistic applications, including roofs, statues, sculptures and façades.

COST AND SPEED Molds and jigs are not required for this process. Cycle time depends on the size and complexity of the work, but is typically around one day. Labour costs are moderate to high due to the level of skill required.

MATERIALS Copper alloys, including copper, brass and bronze, are suitable for artificial patination.

ENVIRONMENTAL IMPACTS Although chemicals are used in the patination process there are no hazardous by-products.

This process protects and passivates the surface against further corrosion, so it lasts longer and requires very little maintenance.

Patinated bronze bust (far left) Sculpted by artist Mark Swan, this bronze bust has been patinated with potassium sulphide to create a gold-brown colour known as 'liver'.

Colour samples (near left) A range of colours and effects is achieved with a range of chemicals: green surface colour is produced using copper nitrate, silver with silver nitrate, red or 'ferric' with iron nitrate, white with titanium or zinc oxide and brown or 'liver' with potassium sulphide.

Carefully darkening the recesses and lightening the peaks enhances the profile and surface texture.

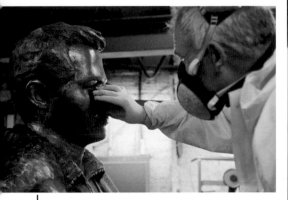

1

2

Case Study

Patinating a Bronze Casting

Featured company Bronze Age Ltd
www.bronzeage.co.uk

This lost wax cast sculpture of Gregoris Afxentiou, 'The Freedom Fighter', was created by Nikolaos Kotziamanis for the Athens War Museum, Greece. The surface is cleaned in preparation for patination (image **1**). A small amount of potassium sulphide (image **2**), mixed with water, is applied to the surface with a stippling brush. This produces a dark brown, natural-looking colour. The chemical reaction is accelerated with a hot flame (image **3**). The surface is sealed with wax for a gloss finish and to protect the colour (image **4**). The finished sculpture in Athens (image **5**).

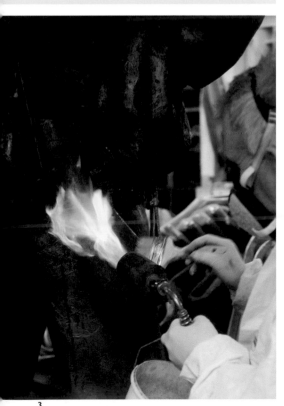

3

4

5

Grinding, Sanding, Polishing

Surfaces are eroded by abrasive particles in these mechanical processes. Surface finish ranges from coarse to mirror and can be a uniform texture or patterned, depending on the technique and type and size of abrasive particle. Grinding, sanding and polishing include a wide range of techniques.

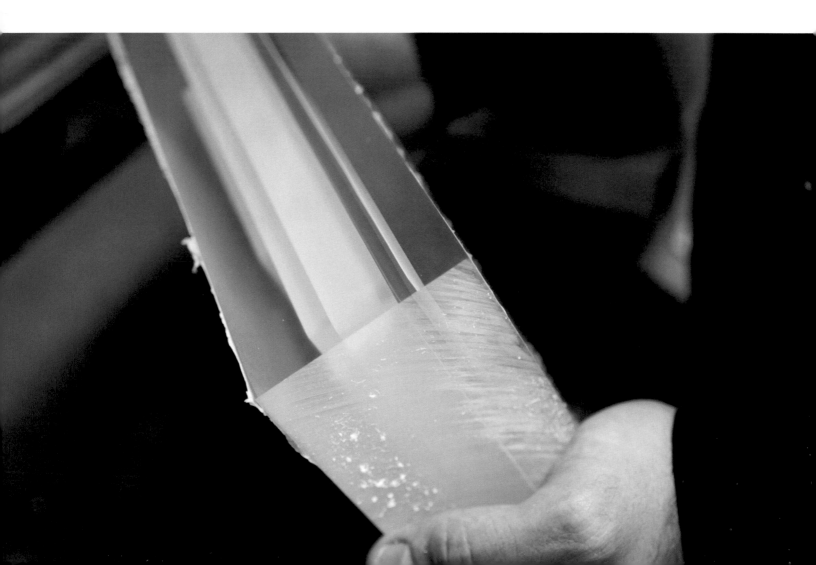

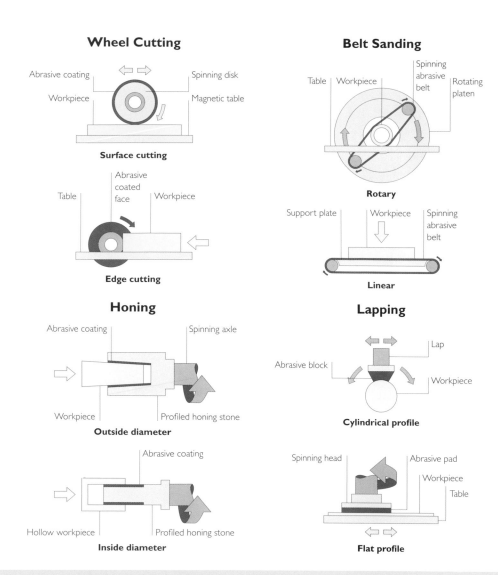

Wheel Cutting

Abrasive coating
Spinning disk
Workpiece
Magnetic table

Surface cutting

Abrasive coated face
Table
Workpiece

Edge cutting

Belt Sanding

Spinning abrasive belt
Table | Workpiece
Rotating platen

Rotary

Support plate | Workpiece | Spinning abrasive belt

Linear

Honing

Abrasive coating
Spinning axle
Workpiece | Profiled honing stone

Outside diameter

Abrasive coating
Hollow workpiece | Profiled honing stone

Inside diameter

Lapping

Abrasive block
Lap
Workpiece

Cylindrical profile

Spinning head | Abrasive pad
Workpiece
Table

Flat profile

Essential Information

VISUAL QUALITY	● ● ● ● ● ○ ○ ○
SPEED	● ● ● ● ● ● ○ ○
MOLD AND JIG COST	● ● ● ● ● ● ○ ○
UNIT COST	● ● ● ● ● ● ○ ○
ENVIRONMENT	● ● ● ● ● ● ● ○

Alternative and competing processes include:
• Metal Patination
• Spray Painting

What is Mechanical Grinding, Sanding and Polishing?

These are common techniques used for cutting surfaces in industrial applications. Each is capable of applying a range of finishes, from super bright to very coarse, depending on the type of abrasive material. They all rely on lubrication, which reduces the build up of heat and wear on the cutting tool.

To achieve a highly reflective and super bright finish, the material will pass through a series of stages of surface cutting, which will use gradually finer grits of abrasive. The role of each abrasive is to reduce the depth of surface undulation. In mirror polishing this is measured in terms of Ra (roughness average); a mirror polish is less than Ra 0.05 microns (0.0000019 in.).

QUALITY It is possible to grind and polish surfaces accurate to within fractions of a micron. Precision operations are considerably more expensive and time consuming, but are sometimes the only viable method of manufacture.

TYPICAL APPLICATIONS These processes are used in all manufacturing industries both for surface preparation and as precise finishing operations. Applications are widespread and cover both industrial and DIY projects.

COST AND SPEED Much grinding, sanding and polishing can be carried out with standard equipment. Consumables are a consideration for the unit price. Cycle time and labour costs are dependent on the size, complexity and smoothness of surface finish. Brightly polished surfaces can take many hours to achieve.

MATERIALS Most materials can be ground, sanded or polished, but the hardness of the material will affect how well it finishes, as well as how long it will take.

ENVIRONMENTAL IMPACTS Even though these are reductive processes, there is very little waste produced in operation.

Vibratory finishing with natural ingredients A press-formed metal part is placed into the vibrating barrel which is filled with smooth, hard pellets (above). It works in much the same way as pebbles eroding each other on the beach to produce smooth, rounded stones.

Crushed maize seed (right), which is used to produce a very fine finish on hard surfaces, is the next stage in the abrasive process.

1

Diamond Wheel Polishing

Featured company Zone Creations
www.zone-creations.co.uk

Diamond particles are used to finish a range of materials that are too hard for other polishing compounds. They are also used for the high-speed finishing of plastics. This acrylic box has been CNC machined (page 42). The lid and base are paired up and then the four sides are cut on a circular saw (image **1**). This produces an accurate finish, but it has an undesirable texture.

The workpiece is placed on the cutting table and clamped in place. The diamond cutting wheel is spinning at high speed and produces a super fine finish in a matter of seconds (images **2** and **3**).

2

3

Case Study

Honing Glass

Featured company Dixon Glass www.dixonglass.co.uk

In this case honing is being used to produce an airtight seal between a glass container and a stopper.

The profiled metal-honing stone is machined specially for this application. It is loaded into the chuck of a lathe and coated with a mineral-based cutting compound (image **1**). As the part is ground, the coarseness of abrasive compound is reduced to produce a finer finish (image **2**). Abrasive honing increases the size of hole very gradually until the glass stopper fits perfectly into it (image **3**). The glass stopper is finished in the same way.

The finished products (image **4**) are used to store biological specimens. Borosilicate glass is the most suitable due to its neutrality and inertness to chemicals over time.

1

2

3

4

Case Study

Mechanical Polishing

Featured company Professional Polishing Services
www.professionalpolishing.co.uk

Lapping is used to produce a range of finishes, including reciprocal patterns, dull and super bright finishes. To polish a sheet of stainless steel to a bright finish, the polishing compound is applied with a roller each time the lapping pad passes over the surface (image **1**). Afterwards a lime substitute is spread over the surface of the stainless steel to remove any remaining moisture (image **2**).

Repeating patterns are produced by circle polishing with circular pads (image **3**) that are impregnated with abrasive particles. The pattern (image **4**) is a typical example of the finish that can be achieved by this method. A variety of different finishes can also be produced.

1

2

3

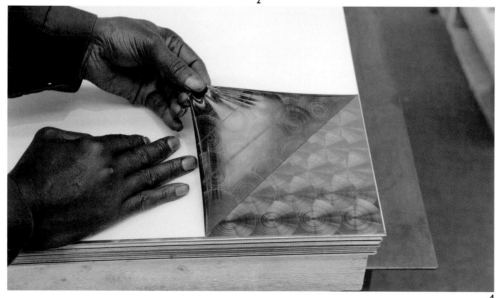

4

This is the most expensive and time-consuming method of polishing. It is used to produce an extremely high-quality surface finish on the surface of parts that cannot be polished by mechanical means. This polishing method can be used to work every surface of this colander (images **1**–**3**). The size of spinning mop can be adjusted to fit smaller profiles such as spoons. The density of the mop is adjusted for different grades of cutting compound.

The final stage is a buffing process, which uses the finest polishing compound to produce a highly reflective finish. It is a labour intensive process, but is nonetheless used to finish a high volume of products such as the Max le Chinois colander designed by Philippe Starck (image **4**).

4

2

3

Case Study

Rotary Belt Sanding

Featured company Pipecraft www.pipecraft.co.uk

This is a typical application for the rotary belt-sanding method (image **1**). It produces an even satin finish on the surface of stainless steel (image **2**). It is equally suitable for non-uniform profiles.

1

2

1

2

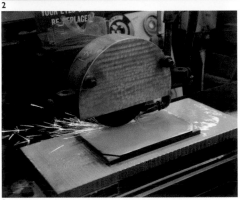

3

4

Wheel Grinding

Featured company CRDM www.crdm.co.uk

Wheel grinding is used to produce a very flat surface that is suitable for rapid prototyping onto using the direct metal laser sintering (DMLS) process (page 108). Firstly, the surface is milled to provide an even surface for grinding (images **1** and **2**). This speeds up the process. Then the metal plate is placed onto the magnetic grinding table and carefully ground for up to 45 minutes to produce the desired Ra value (images **3** and **4**).

Spray Painting

Spray painting is used to apply colour and provide surface protection. It can be used to coat most rigid materials. A range of finishes can be achieved from matt to high gloss, including lustrous colour, soft touch (also known as warm to touch), pearlescent, iridescent and metallic.

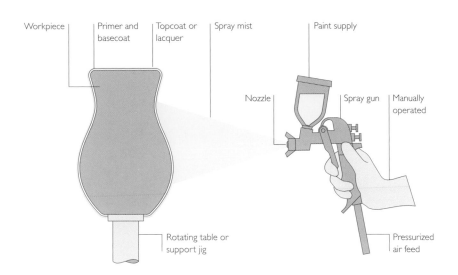

Workpiece | Primer and basecoat | Topcoat or lacquer | Spray mist | Paint supply

Nozzle | Spray gun | Manually operated

Rotating table or support jig | Pressurized air feed

Essential Information

VISUAL QUALITY	○○○○○○●●
SPEED	○○○●●●●●
MOLD AND JIG COST	○○●●●●●●
UNIT COST	○○○○○●●●
ENVIRONMENT	○○○○●●●●

Alternative and competing processes include:
- Dip Molding
- Electroplating
- Flocking
- Metal Patination
- Polishing
- Vacuum Metalizing

What is Spray Painting?

Spray guns use a jet of compressed air to atomize the paint into a fine mist. The atomized paint is blown out of the nozzle in an elliptical shape. The coating is applied onto the surface in overlapping strokes.

Painted surfaces are nearly always made up of more than one layer. If necessary, the surface is prepared with filler and primer.

Conventional paints are made up of pigment, binder, thinner and additives. The role of the binder is to bond the pigment to the surface being coated. It determines

the durability, finish, speed of drying and resistance to abrasion. These mixtures are dissolved or dispersed in either water or a solvent.

Two-pack paints are made up of resin and the catalyst or hardener. They bond to the surface in a one-way reaction and are extremely durable.

QUALITY The level of sheen on the coating is categorized as matt (also known as egg shell), semi-gloss, satin (also known as silk) and gloss. High gloss, intense and colourful finishes are produced by a combination of meticulous surface preparation, basecoat and topcoat.

TYPICAL APPLICATIONS Spray painting is used in a vast range of applications including prototyping, low-volume and high-volume production.

COST AND SPEED Jigs may be required, but this depends on the number of parts and geometry. Cycle time is rapid, but depends on the size, complexity, number of coats and drying time. Labour costs are typically high because these tend to be manual processes.

MATERIALS Almost all materials can be spray painted. Some surfaces have to be coated with an intermediate layer which is compatible with both the workpiece and the topcoat.

ENVIRONMENTAL IMPACTS Water-based paints are less toxic than solvent borne. Spraying is usually carried out in a booth or cabinet to allow the paints to be recycled and disposed of safely.

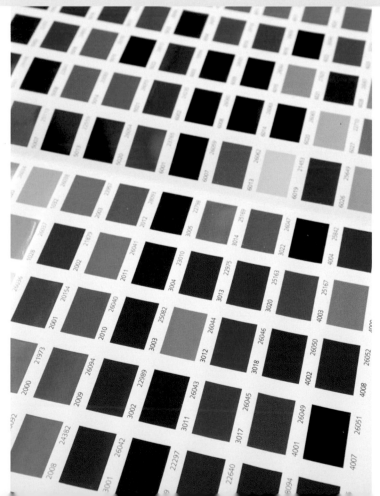

Outrageous colour shifting paint (above) The colour of this paint changes dramatically depending on the viewing angle. It is supplied by Cambridgeshire Coatings in the UK and US Chemicals and Plastics in the US.

RAL colour reference chart (left) There is an almost unlimited range of colours and finishes for paints. Standard colour ranges include RAL (German origin) and Pantone (US origin).

Colour is supplied by pigments, which are solid particles of coloured material. They can be replaced or enhanced by platelets of metallic, pearlescent, dichroic, thermochromic or photoluminescent materials.

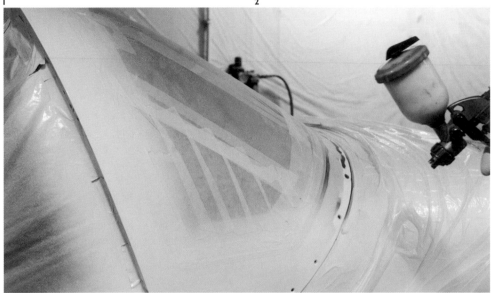

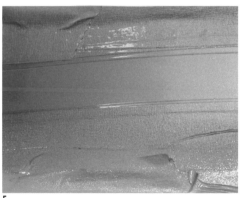

1

2

3

4

5

Spray Painting a Pioneer 300 Light Aircraft

Featured company Hydrographics
www.hydro-graphics.co.uk

This is the Pioneer 300 (image **1**) manufactured by Alpi Aviation, Italy. The joints are filled and rubbed down to make a smooth surface for the primer. Masking is used to protect the glass hood and areas of the bodywork (image **2**).

The first coat of polyurethane two-pack primer is sprayed on in overlapping strokes (image **3**). Three coats are applied in total. Once the primer is sufficiently dry, the fuselage is rubbed down with abrasive paper to produce a smoother finish.

In preparation for the topcoat, the top layer of masking tape is removed (image **4**). Underneath is a second layer of masking which marks where the topcoat will be sprayed up to. The edges are staggered in this way to avoid a build up of paint; otherwise the primer would produce a visible white line along the masked edge. Finally, the topcoat is applied, including the bright blue 'racing' stripe (image **5**).

Case Study

High Gloss Black Lacquered Yamaha Piano

Featured company Yamaha Corporation
www.yamaha.com

High gloss finishes are the result of meticulous surface preparation, spraying, sanding and polishing. This high gloss black lacquered piano, from Yamaha, can take up to a week to prepare and finish (image **1**).

Firstly, the panels to be coated are laminated with a particle board, which provides the optimum base for spray painting. The panels are coated four times in quick succession with a high-build polyester lacquer (image **2**). This provides a sufficient thickness for sanding and polishing.

The piano bodywork is air-dried for three days to allow the polyester to cure fully (image **3**). Each part is very carefully sanded by hand, using blocks and sanding belts to create a very even surface finish (image **4**).

A mirror finish is achieved by patiently progressing through the grades of abrasives and polishing compounds (image **5**). All parts are finished by hand to ensure the highest level of surface quality (image **6**).

The pianos are assembled on a slow moving production line (image **7**). As acoustic instruments, the construction has to be extremely precise. Skilled craftsmanship and a great deal of work results in a superior product (image **8**).

5

6

7

8

Glossary

CAD

Computer-aided design (CAD) is a general term used to cover computer programmes that assist with engineering and product design. Some of the most popular 3D packages include Professional Engineer, commonly known as Pro E, Rhino, SolidWorks, AutoCAD, Alias and Maya.

CNC

Machining equipment that is operated by a computer is known as computer numerical control (CNC). The number of operational axes determines the types of geometries that can be achieved. Two axis, three axis and five axis are the most common (page 42).

Direct manufacturing

Manufacturing products directly from CAD data such as rapid prototyping (page 104).

Elastomer

A natural or synthetic material that exhibits elastic properties: the ability to deform under load and return to its original shape once the load is removed.

Ferrous

Metals that contain iron such as steel. See also non-ferrous.

FRP

Molded plastic reinforced with lengths of fibre – including carbon, aramid, glass or natural material such as cotton, hemp or jute – is known as fibre-reinforced plastic (FRP). These materials are utilized in composite laminating (page 94). See also GRP.

GRP

Molded plastic reinforced with lengths of glass fibre is known as glass-reinforced plastic (GRP). Composite laminating (page 94) uses long or continuous lengths of glass fibre reinforcement. See also FRP.

Hardwood

Wood from deciduous and broad-leaved trees such as birch, beech, ash and oak.

Mold

A hollow form, which is used to shape materials in their liquid or plastic state, or a concave or convex 3D profile over which materials are formed in their solid state.

Monomer

A small, simple compound that can be joined with other similar compounds to form long chains known as polymers.

Non-ferrous

Metals that do not contain iron such as aluminium alloys and copper alloys. See also ferrous.

Pattern

An original design or prototype that is reproduced to form a mold, such as in composite laminating (page 94) or investment casting (page 30). This mold can then be used to produce many identical parts.

Polymer

A natural or synthetic compound made up of long chains of repeating identical monomers.

Ra

Surface texture on ground, sanded and polished surfaces is measured as a value of roughness average (Ra). Approximate Ra values for metal finishes are as follows: a ground finish obtained with 80–100 grit will be Ra 2.5 microns (0.000098 in.) and a bright polish produced with a polishing compound will be Ra 0.05 microns (0.0000019 in.).

RAL

Reichsausschuss für Lieferbedingungen is a German colour chart system used mainly in paint and pigment colour specification.

Resin

A natural or synthetic semi-solid or solid substance, produced by polymerization or extraction from plants, and used in plastics, varnishes and paints.

RTV

Certain rubbers, such as silicone, can be chemically cured at room temperature, as opposed to heat curing, in a process known as room temperature vulcanizing (RTV).

Shore hardness

The hardness of a plastic, rubber or elastomer is

measured by depth of indentation of a shaped metal foot on a measuring instrument known as a durometer. The depth of indentation is measured on a scale of 0 to 100; higher numbers indicate harder materials. These tests are generally used to indicate the flexibility of a material. The two most popular are Shore A and Shore D. There is not a strong correlation between different scales.

Softwood
Wood from coniferous and typical evergreen trees such as pine, spruce, fir and cedar.

Thermoplastic
A polymeric material that becomes soft and pliable when heated.

Thermosetting plastic
A material formed by heating, catalyzing, or mixing two parts to trigger a one-way polymeric reaction. Thermosetting plastics form cross links between the polymer chains, which cannot be undone, and so this material cannot be re-shaped or re-molded once cured.

Tool
This is another term for mold.

Two-pack or two-part
These terms are used to describe the two-part nature of thermosetting plastics or rubbers, such as those used in spray painting (page 180) and reaction injection molding (page 14).

VDI
Surface texture is measured on the Association of German Engineers' VDI scale. The VDI scale is comparable with roughness average (Ra) 0.32–18 microns (0.000013–0.00071 in.).

Featured Companies

Alessi SpA
Via Privata Alessi, 6
28887 Crusinallo Di
Omegna – VB
Italy
www.alessi.com

Aspect Signs & Engraving
Unit A5, Bounds Green
Industrial Estate
Ringway
London N11 2UD
United Kingdom
www.aspectsigns.com

Bakers Patterns Ltd
F5 Stafford Park 15
Telford TF3 3BB
United Kingdom
www.polystyrenemodels.co.uk

Benjamin Hubert Studio
Unit AH 220, Aberdeen Centre
22–24 Highbury Grove
London N5 2EA
United Kingdom
www.benjaminhubert.co.uk

BJS Company
65 Bideford Avenue
Perivale, Greenford
Middlesex UB6 7PP
United Kingdom
www.bjsco.com

Bronze Age Ltd
272 Island Row
Limehouse
London E14 7HY
United Kingdom
www.bronzeage.co.uk

Cambridgeshire Coatings
PO Box 34
St Neots
Huntingdon PE19 6ZG
United Kingdom
www.outrageousfinishes.co.uk

Cathryn Shilling Glass
7 Mortimer Road
London W13 8NG
United Kingdom
www.cathrynshilling.co.uk

Chiltern Casting Company
Unit F, Cradock Road
Luton LU4 0JF
United Kingdom

CMA Moldform Limited
Unit 17A, Spitfire Park
Spitfire Road
Birmingham B24 9PR
United Kingdom
www.cmamoldform.co.uk

Cove Industries
Industries House
18 Invincible Road
Farnborough
Hampshire GU14 7QU
United Kingdom
www.cove-industries.co.uk

CPP (Manufacturing) Ltd
Wheler Road
Seven Starts Industrial Estate
Coventry
West Midlands CV3 4LB
United Kingdom
www.cpp-uk.com

CRDM
Wycombe Sands
Lane End Road
High Wycombe
Buckinghamshire HP12 4HH
United Kingdom
www.crdm.co.uk

Crompton Technology Group
Thorpe Park
Thorpe Way
Banbury
Oxfordshire OX16 4SU
United Kingdom
www.ctgltd.co.uk

Daniel Rohr
Bleichstr. 8
64283 Darmstadt
Germany
www.daniel-rohr.com

Dixon Glass
127–129 Avenue Road
Beckenham
Kent BR3 4RX
United Kingdom
www.dixonglass.co.uk

Droog Design
Staalstraat 7A–7B
1011 JJ Amsterdam
The Netherlands
www.droog.com

FIAM Italia SpA
Via Ancona 1
61010 Tavullia – PU
Italy
www.fiamitalia.it

**Hartley Greens & Co.
(Leeds Pottery)**
Anchor Road
Longton
Stoke-on-Trent ST3 5ER
United Kingdom
www.hartleygreens.com

Hydrographics
Unit 4, Brockett Industrial Estate
York YO23 2PT
United Kingdom
www.hydro-graphics.co.uk

Hymid Multi-Shot Ltd
12 Brixham Enterprise Estate
Rea Barn Road
Brixham TQ5 9DF
United Kingdom
www.hymid.co.uk

Instrument Glasses
236–258 Alma Road
Ponders End
Enfield EN3 7BB
United Kingdom
www.instrumentglasses.com

Isokon Plus
Windmill Furniture
Turnham Green Terrace Mews
London W4 1QU
United Kingdom
www.isokonplus.com

Lola Cars International
Glebe Road
St Peters Road
Huntington
Cambridgeshire PE29 7DS
United Kingdom
www.lolacars.com

London Glassblowing
62–66 Bermondsey Street
London SE1 3UD
United Kingdom
www.londonglassblowing.co.uk

Midas Pattern Co. Ltd
22 Shuttleworth Road
Bedford MK41 0RX
United Kingdom
www.midas-pattern.co.uk

MiMtec Limited
Unit D, Jten Trade Park
Wickham Road
Fareham
Hampshire PO16 7JB
United Kingdom
www.mimtec.co.uk

National Glass Centre
Liberty Way
Sunderland SR6 0GL
United Kingdom
www.nationalglasscentre.com

Portable Welders
2 Wedgwood Road
Bicester OX26 4UL
United Kingdom
www.portablewelders.com

Professional Polishing Services
18B Parkrose Industrial Estate
Middlemore Road
West Midlands B66 2DZ
United Kingdom
www.professionalpolishing.co.uk

Rachel Dormor Ceramics
Unit 1, Dwight Creative Village
Leek Road Business Village
Stoke-on-Trent
Staffordshire ST4 2AR
United Kingdom
www.racheldormorceramics.com

Radcor
Hingham Road Industrial Estate
Great Ellingham
Norfolk NR17 1JE
United Kingdom

Thomas & Vines Ltd
Units 5–6, Sutherland Court
Moor Park Industrial Centre
Tolpits Lane
Watford WD18 9SP
United Kingdom
www.flocking.co.uk

TWI
Granta Park
Great Abington
Cambridge CB1 6AL
United Kingdom
www.twi.co.uk

US Chemical & Plastics
PO Box 709
Massillon, OH 44648
USA
www.uschem.com

Windmill Furniture
Turnham Green Terrace Mews
London W4 1QU
United Kingdom
www.windmillfurniture.com

Yamaha Corporation
10–1 Nakazawa-cho.
Shizuoka Pref.
430-8650 Hamamatsu
Japan
www.yamaha.com

Zone Creations
64 Windsor Avenue
London SW19 2RR
United Kingdom
www.zone-creations.co.uk

Further Reading

Ashby, Mike, and Kara Johnson, *Materials and Design: The Art and Science of Material Selection in Product Design* (Oxford and Boston: Butterworth-Heinemann, 2002)

Beylerian, George M., Andrew Dent and Anita Moryadas (eds), *Material ConneXion: The Global Resource of New and Innovative Materials for Architects, Artists, and Designers* (London: Thames & Hudson and Hoboken, NJ: J. Wiley, 2005)

Brownell, Blaine (ed.), *Transmaterial: A Catalog of Materials that Redefine Our Physical Environment* (New York: Princeton Architectural Press, 2006)

Byars, Mel, *50 Chairs: Innovations in Design and Materials* (Crans-Près-Céligny: RotoVision and New York: Watson-Guptill Publishers, 1996)

Byars, Mel, *50 Lights: Innovations in Design and Materials* (Crans-Près-Céligny: RotoVision, 1997)

Byars, Mel, *50 Products: Innovations in Design and Materials* (Crans-Près-Céligny: RotoVision, 1998)

Byars, Mel, *50 Tables: Innovations in Design and Materials* (Crans-Près-Céligny: RotoVision, 1998)

Chua, C. K., K. F. Leong and C. S. Lim, *Rapid Prototyping: Principles and Applications* (Singapore: World Scientific Publishing, 2003)

Fournier, Ron and Sue Fournier, *Sheet Metal Handbook* (Los Angeles: HP Books, 1989)

Hopkinson, Neil, Richard Hague and Philip Dickens, *Rapid Manufacturing: An Industrial Revolution for the Digital Age* (Hoboken, NJ: Wiley-Blackwell, 2005)

Hudson, Jennifer, *Process: 50 Product Designs from Concept to Manufacture* (London: Laurence King 2008)

IDSA (Industrial Designers Society of America), *Design Secrets. Products: 50 Real-Life Projects Uncovered* (Massachusetts: Rockport Publishers, 2003)

IDTC (International Design Trend Centre), *How Things Are Made: Manufacturing Guide for Designer* (Seoul: Agbook, 2003)

Joyce, Ernest, *The Technique of Furniture Making*, 4th edn, revised by Alan Peters (London: Batsford, 2002)

Kula, Daniel, Elodie Turnaux and Quentin Hirsinger, *Materiology: The Creative Industry's Guide to Materials and Technologies* (Boston, Mass.: Birkhäuser, 2008)

Lefteri, Chris, *Making It: Manufacturing Techniques for Product Design* (London: Laurence King, 2007)

Lefteri, Chris, *Materials for Inspirational Design* (Mies: RotoVision, 2006)

Lesko, Jim, *Industrial Design: Materials and Manufacturing Guide* (Hoboken, NJ: J. Wiley, 1999)

Mori, Toshiko (ed.), *Immaterial/ Ultramaterial: Architecture, Design, and Materials* (Cambridge, Mass.: Harvard Design School in association with George Braziller, 2002)

Saville, Laurel, and Brooke Stoddard, *Design Secrets. Furniture: 50 Real-Life Projects Uncovered* (Massachusetts: Rockport Publishers, 2008

Stattmann, Nicola, *Ultra Light Super Strong: A New Generation of Design Materials* (Basel and Boston: Birkhäuser, 2003)

van Onna, Edwin, *Material World: Innovative Structures and Finishes for Interiors* (Amsterdam/Basel and Boston: Frame Publishers/Birkhäuser, 2003)

Illustration Credits

Rob Thompson photographed the processes, materials and products in this book. The author would like to acknowledge the following for permission to reproduce photographs and CAD visuals.

Prelims
Page 3 (top left): Ester Segarra for Cathryn Shilling Glass

Introduction
Page 8 (Maverick Television Awards): CMA Moldform
Page 9 (Heavy Light): Benjamin Hubert Studio
Page 9 (Five-axis CNC Machining): Bakers Patterns
Page 10 (Signature vase): Droog Design
Page 11 (Blown glass): London Glassblowing

Panel Beating
Page 40 (Austin-Healey 3000): CPP (Manufacturing) Ltd
Page 41 (image 1): Spyker Cars

CNC Machining
Page 44 (A Sculpture by Zaha Hadid Architects): A M Structures
Page 45 (image 1): Bakers Patterns
Pages 46–47 (all images): Daniel Rohr

Ceramic Wheel Throwing
Page 64 (all images): Rachel Dormor Ceramics
Page 65 (image 4): Rachel Dormor Ceramics

Kiln Forming Glass
Page 70 (title images): Ester Segarra for Cathryn Shilling Glass
Page 72 (all images): Ester Segarra for Cathryn Shilling Glass
Page 74 (Textured Glass and Ghost Chair): FIAM Italia
Page 75 (all images): FIAM Italia

Studio Glassblowing
Page 80 (Wide Stone from the Ariel range): London Glassblowing

Veneer Laminating
Page 90 (Cold pressing the Isokon Long Chair, left): Isokon Plus
Page 91 (image 1): Isokon Plus
Page 93 (image 1): Isokon Plus

Composite Laminating
Page 94 (title image): Lola Cars International
Page 96 (Lola B05/30 Formula 3 racing car): Lola Cars International
Page 97 (images 1 and 5): Lola Cars International
Page 99 (image 1): Ansel Thompson

Filament Winding
Page 102 (Encapsulating a Metal Liner with Carbon Fibre Composite, right): Crompton Technology Group

Rapid Prototyping
Page 106 (all images): CRDM
Page 108 (Functional prototypes): CRDM
Page 109 (all images): CRDM
Page 110 (Functional prototypes): CRDM

Arc Welding
Pages 134–141 (all images): TWI

Soldering and Brazing
Page 144 (image 1): Alessi

Resistance Welding
Page 148 (Spot welding stainless steel mesh, right): Portable Welders

Wood Joinery
Page 153 (image 1): Isokon Plus
Page 155 (Housing Joints in the Donkey, image 1): Isokon Plus

Grinding, Sanding, Polishing
Page 177 (image 4): Alessi

Spray Painting
Page 182 (Outrageous Colour Shifting Paint): Cambridgeshire Coatings/ US Chemical and Plastics
Page 183 (image 1): Duncan Cubitt (Today's Pilot)
Page 183 (image 5): Hydrographics
Page 184 (images 1 and 8): Yamaha Corporation

Acknowledgments

The technical detail and accuracy of the book's content is the result of the extraordinary generosity of many individuals and organizations. Their knowledge of processes and materials and their years of hands on experience were fundamental in putting this book together. Thank you to all of the companies that have been featured. Their contact details are given on pages 187 to 189.

I would like to thank the book's designer Chris Perkins, the editor Ilona de Nemethy Sanigar and Thames & Hudson for their continued support and dedication to producing the highest standard of work.

The support, encouragement and input from colleagues, family and friends has been invaluable throughout.

This book is dedicated to Mum, Dad, Ansel, Murray and Lina.

Index